RADIATION SAFETY IN SHELTERS

FOREWORD

Exposure to high levels of nuclear radiation can cause sickness or death. Because people cannot see, hear, smell, taste, or feel nuclear radiation, they are unable to tell without the proper radiological instruments whether the levels of radiation around them might be harmful.

This handbook is written for radiation safety in shelters in areas that will not be affected by the primary nuclear weapons effects of blast, fire, and initial nuclear radiation. In a nuclear war, up to 90 percent of the land area of the 48 states of the United States could be covered with radioactive fallout that would deliver hazardous nuclear radiation to an unprotected person over a period of several days before decaying to much less hazardous levels. On the other hand, about 10-15 percent of the land area could be affected by primary nuclear weapons effects that would pose additional hazards to the population remaining there. In those areas, additional safety measures must be taken that are not described in this handbook. Nearly all the radiation safety measures and procedures described in this handbook will be useful in all shelters. The procedures for watching for fallout to arrive should not be followed in shelters that are less than 25 miles from a likely target for a nuclear weapon. At such locations, the possibility of other nuclear weapons effects such as blast and thermal radiation will place the Radiological Monitor (RM) following this procedure under greater risk than necessary.

Louis O. Giuffrida
Director

ISBN: 978-1-60322-095-8

TABLE OF CONTENTS

TABLE OF CONTENTS (continued)

TABLE OF CONTENTS (continued)

LIST OF FIGURES

LIST OF TABLES

CHECKLIST "A"

(FOR IMMEDIATE ACTION)

STEP 1. Select Radiological Monitors (RMs) for Your Shelter. If you have already been selected to be the RM, go on to STEP 2.

Read paragraph 1-2, page 1-2, of the handbook on selecting RMs. Give this copy of Radiation Safety in Shelters to a person selected to be an RM, unless he or she already has a copy and there are enough copies for all the radiological monitors. Keep it for your own use if you are selected or were already appointed to be an RM for your shelter.

STEP 2. Get Radiological Instruments and Forms for Radiation Exposure Records for Your Shelter. If you already have these items, go on to STEP 3, although you may wish to read paragraphs 2b and 2c below to check whether you have the right kind of instruments and forms and whether you have enough of them.

a. Where

The instruments and forms may already be in your shelter or may be in the process of being delivered to your shelter, or you may have to go and get them. You will have to find out what the situation is if you don't already know. Listen to your radio for information. If you don't have instruments or forms at the shelter, try to contact your local government, city or county, about getting them.

If you don't have a telephone or a two-way radio in your shelter or if the lines are jammed (don't spend more time trying to place your call than one-third of the time it would take you to actually get there), it may be necessary for you to get in a car and drive to your local city hall or county seat and find out about the radiological instruments. Consult your Shelter Manager about the advisability of going. Don't go if fallout has already arrived, or if you expect it to arrive before you return. You may be able to detect the arrival of fallout without having radiological instruments. For information on the arrival of fallout, you may wish to read, if you have time, paragraph 1-10, page 1-12, of the handbook.

If no radiological instruments are available for your shelter, try to locate the best shelter available. See paragraph 4-2b(1), page 4-5, of the handbook for ideas on getting the best protection within the shelter.

CHECKLIST "A" (continued)

b. What Kind

Three types of instruments are required:

(1) survey meter (photograph shown in Figure 2-1, page 2-2);
(2) dosimeter (Figure 2-2, page 2-3); and
(3) charger (Figure 2-3, page 2-5).

The survey meter and the charger each require one D-cell battery, a battery commonly used in flashlights.

The form for keeping records of radiation exposure is shown in Figure 4-1, page 4-3. A few forms are at the back of this handbook. If necessary, additional forms can be duplicated, or they can be made by hand on any available writing surface.

c. How Many

(1) Survey meters. It would be desirable to have at least one survey meter if your shelter has up to 200 occupants; two survey meters (and two RMs for each shift) if your shelter has 200 to 400 occupants; and so on. You may get less because there may not be enough instruments.

(2) Dosimeters. It would be desirable to have at least one dosimeter for every 10 occupants of your shelter, plus additional dosimeters for RMs and shelter managers. Again, you may get less because there may not be enough instruments.

(3) Chargers. One charger can service many dosimeters. It would be desirable to have about as many chargers as you have survey meters.

(4) Batteries. You should have one extra D-cell battery for each instrument that uses a battery.

(5) Forms for radiation exposure records. You should have one per shelter occupant, plus about 5 percent more to allow for errors and accidents.

STEP 3. Check the Instruments. If you have already checked the instruments, go on to STEP 4.

If possible, get instructions and a demonstration on checking and operating the instruments from someone who knows how, perhaps from the person who delivers the instruments to your shelter or from the person who hands you the instruments at the warehouse or headquarters of your local government. IF YOU CAN'T GET INSTRUCTIONS, follow this procedure:

CHECKLIST "A" (continued)

a. Install a D-cell battery in the survey meter if it doesn't have one already. It should not be stored with a battery in place, but one may have been installed in the process of issuing the instruments. First, read paragraph 3-2b, page 3-1, of the handbook and then follow the instructions.

b. Perform an operational check on the survey meter. First, read paragraph 3-2c, page 3-2, of the handbook and then follow the instructions.

c. Install a D-cell battery in the charger, if it doesn't have one already. First, read paragraph 3-3b, page 3-8, of the handbook and then follow the instructions.

d. Perform an operational check on the charger. First, read paragraph 3-3c, page 3-8, of the handbook and then follow the instructions.

e. Charge (zero) the dosimeters. First, read paragraph 3-4b, page 3-13, of the handbook and then follow the instructions. These instructions are the same as for the operational check of the charger, except for step 7 in paragraph 3-4b, page 3-14.

STEP 4. Give Out the Dosimeters and Radiation Exposure Record Forms. Keep one dosimeter and the extra Radiation Exposure Record forms for yourself. Each Radiological Monitor, the Shelter Manager and his assistants, and each Unit Leader should get a dosimeter. List the serial number of each dosimeter, print the name of the person to whom you issue it next to the serial number, and have the person sign his or her name to acknowledge receipt of the dosimeter. They should be instructed to wear them either in a breast pocket or clipped to the collar, neckline, or belt. Each Unit Leader should get one Radiation Exposure Record form for each person in his or her unit. At this time, tell the Unit Leaders that you will tell them later what they are supposed to do with the dosimeters and forms. When you have time, read paragraphs 4-2a, page 4-2, and 4-4f, page 4-32, of the handbook to find out what the Unit Leaders are supposed to do.

STEP 5. What You Do Next Depends on the Situation. One of the following situations may apply:

a. If fallout has already started to come down around your shelter and people are still arriving, you will need to set up some procedures for decontamination of these people. Decontamination procedures are described in paragraph 4-4a, page 4-19, of the handbook. Do not allow this process to cause a blockage of the entranceway that would result in having people stand outside in the fallout.

CHECKLIST "A" (continued)

b. If fallout has already arrived at your shelter and no more people are coming to your shelter, prepare to find the places with the lowest radiation levels in your shelter. If you don't know what this requires, take time to read through paragraphs 4-4b, page 4-20, and 4-4c, page 4-22, of the handbook.

c. If fallout has not yet arrived at your shelter but is thought to be on the way, based on radio announcements or other signs and indicators, you should set up a watch for the arrival of fallout. Read paragraph 4-3, page 4-16, of the handbook. While you are waiting for fallout to arrive, you should study paragraph 4-4, page 4-19, so you will know what to do after fallout arrives. Then, if you still have time, you should study Chapter 1, which gives some facts about nuclear radiation.

d. If a nuclear attack has not occurred but may possibly start in the next few hours, you may have time to check out the shelter and make some improvements in its radiation safety. Read paragraph 4-2b, page 4-4, of the handbook. If you have time, study Chapter 1 also.

CHECKLIST "B"

(STANDARD CHECKLIST FOR RADIOLOGICAL MONITORS)

This checklist may be used during a crisis period when a nuclear attack is not expected for many hours or several days. It may also be used for training purposes. It is assumed that you, the RM, have studied this handbook and have been assigned to be the RM for a particular shelter in your community. If the people in your community are to be relocated in the event of a severe international crisis, you may also be assigned to be the RM for another particular shelter in the relocation area. It is also assumed that radiological instruments are at your shelter when you begin this checklist.

STEP 1. Check the Instruments.

a. Install a D-cell battery in the survey meter, if it doesn't have one already (paragraph 3-2b, page 3-1, of the handbook). It should not be stored with a battery in place, but one may have been installed in the process of issuing the instruments.

b. Perform an operational check on the survey meter (paragraph 3-2c, page 3-2).

c. Install a D-cell battery in the charger, if it doesn't have one already (paragraph 3-3b, page 3-8).

d. Perform an operational check on the charger (paragraph 3-3c, page 3-8).

e. Charge (zero) the dosimeters (paragraph 3-4b, page 3-13). Record the serial number for each dosimeter and the time at which you zero it. Leave room on the list for the name of the person to whom the dosimeter will be issued.

STEP 2. Give Out the Dosimeters and Radiation Exposure Record Forms. Keep one dosimeter and extra Radiation Exposure Record forms for yourself. Issue one dosimeter to the Shelter Manager, and to each shelter manager assistant, RM, and Unit Leader. Put their names on the list by the serial numbers of the dosimeters, and have them sign the list to acknowledge receipt of the dosimeter. Issue one Radiation Exposure Record form to each of these people, except to the Unit Leader, who will need one form for each member of his or her unit.

STEP 3. Give Instructions to Those People Given Dosimeters.

a. Show them how to read the dosimeters. Follow the instructions in paragraph 3-4d, page 3-16.

CHECKLIST "B" (continued)

b. Show them how to wear the dosimeters (paragraph 2-4, page 2-3).

c. Have them copy the three radiation sensitivity categories listed in Table 4-1, page 4-4, so they will know what they are and will be able to help others fill out their forms properly.

d. Show them how the Radiation Exposure Record will be filled out. Use the sample in Figure 4-8, page 4-35. The first entry will show the time at which fallout begins, in the "Comments" column. Inform the Unit Leaders that they will be responsible for estimating the radiation exposure of each individual in their units, based on the dosimeter readings on their own dosimeters, with additional exposure estimated for individuals who make special trips that subject them to greater radiation exposure.

e. Request that they all return in 24 hours, if convenient, to the same location where the dosimeters were issued so the dosimeters can be checked for leaks (paragraph 3-4c, page 3-15). If a dosimeter shows a reading of 2 to 3 R before the 24 hours are up and no fallout has arrived, that dosimeter should be brought back to you so you can try to get a replacement. If no replacements are available, you can still use the dosimeter, as described in paragraph 3-4c, page 3-15.

STEP 4. Secure the Survey Meter and Charger. Find a place where you can lock up these instruments and leave them there, or else leave them with another RM, while they are not needed. Take the batteries out whenever you store the instruments.

STEP 5. Get a Sketch of Your Shelter Floor Plan. If your Shelter Manager doesn't have a sketch of your shelter floor plan, you should make a rough sketch or have someone among the occupants make one for you. See Figure 4-3, page 4-8, for a sample sketch of a floor plan. You should walk through the entire shelter before you begin sketching. If people are gathering in the shelter and beginning to set up housekeeping, it will be helpful if you wear some type of identification, such as an armband with the initials "RM," so people will know who you are.

STEP 6. Locate the Areas that Appear to Provide the Best Protection Against Fallout (paragraph 4-2b(1), page 4-5). Have a discussion with the shelter manager about these areas.

STEP 7. Estimate Whether There Will Be Enough Room in the Locations of Best Protection (paragraph 4-2b(2), page 4-8). Discuss this situation with the Shelter Manager.

CHECKLIST "B" (continued)

STEP 8. Look for Ways to Improve the Shielding of the Shelter (paragraph 4-2b(3), page 4-10). If you think significant improvements can be made with the materials, manpower, tools, and time available, discuss your plan with the Shelter Manager. In some communities, there may be detailed plans already made for upgrading your shelter during a crisis period.

STEP 9. Check for Openings that Might Provide a "Leak" for Gamma Radiation, or Might Let the Wind Blow Fallout Into the Shelter (paragraph 4-2b(4), page 4-10). Remember that in a crowded shelter you will need much more ventilation than you would ordinarily need. You should discuss your plans with the Shelter Manager before he assigns a work crew to cover up any openings.

STEP 10. Locate Materials and Tools that Might Possibly be Used for Improvising Shielding after Fallout Arrives (paragraph 4-2b(5), page 4-12.).

STEP 11. Check the Entranceways for Possible Traffic Problems (paragraph 4-2b(6), page 4-12). If the shelter is inside a large building, there should be signs showing people where to go. You may need to set up receptionists at the entrances. If you think this is necessary, discuss the situation with the Shelter Manager and let him select people to be receptionists. You will need to tell them what to do.

STEP 12. Locate Water, Food Supplies, and Restrooms (paragraph 4-2b(7), page 4-13). Estimate whether trips for supplies or to the restrooms will require extra monitoring for radiation exposure. Check whether there is a possibility of fallout getting into your water supply. If there is such a possibility, see if the water supply can be covered. If not, you may need a supply of potassium iodide (KI) tablets to provide blocking doses to prevent the possibility of radioiodine concentrating in the thyroids of those drinking the water (paragraph 1-8e, page 1-9). These tablets may be obtained at some drugstores, if they are not made available through your local or state government.

STEP 13. Find Locations Where Dosimeters Could be Hung or Mounted (paragraph 4-2b(8), page 4-13). At certain times, while sleeping, for example, the Unit Leaders will need to hang or mount their dosimeters in the general vicinity of their units. It would be handy to have string, tape, and thumbtacks for this purpose.

STEP 14. Make Sure You Have a Reliable Light Source (paragraph 4-2b(10), page 4-14).

CHAPTER 1

GENERAL INFORMATION

1-1. Why You Need This Handbook. High levels of nuclear radiation can make you sick or kill you. You can detect nuclear radiation with the right kind of instruments. Because you cannot see, hear, smell, taste, or feel radiation, you will not be able to tell without these instruments whether the levels of nuclear radiation around you can make you sick.

a. This handbook will give you some facts about nuclear radiation, such as:

(1) What it is;

(2) How it is produced;

(3) How it can make you sick or kill you;

(4) Why you can't feel it;

(5) How it is measured;

(6) How much is harmful;

(7) How you can shield yourself from it; and

(8) How it will fade away.

b. If you have to assume the responsibilities of Radiological Monitor for your shelter without previous training, this handbook will tell you HOW TO USE INSTRUMENTS so you can:

(1) Detect nuclear radiation;

(2) Find the places with the lowest nuclear radiation levels in a shelter;

(3) Improve the protection of places with the lowest nuclear radiation levels so the radiation is reduced even more;

(4) Advise when (and for how long) someone can go outside the shelter on short emergency trips; and

(5) Advise when to leave for longer trips and when to leave permanently.

c. If you have had training in radiological monitoring, this handbook will be useful as a reference.

1-2. Selection of Radiological Monitors. The selection of Radiological Monitors (RMs) for your shelter should have been made by the county or local government. If a nuclear war emergency should arise, and no RMs have been selected for your shelter, or if the selected RMs are not able to get to the shelter in time, then you will need to select RMs from the men and women who have assembled at your shelter. RMs may be selected by a group gathered together of all those who have had technical training or experience and have worked with instruments. Anyone who has studied this handbook is qualified and should volunteer to be a member of the group that selects RMs for a shelter.

The number of RMs you should have in your shelter will depend on how many people are in your shelter, how many survey meters you have, and whether your shelter has unusual radiation safety problems. If there are many people in your shelter, there may be more than one survey meter in it. There should be enough RMs in a shelter to provide round-the-clock (24-hour) radiological monitoring with the available survey meters. Monitoring may be organized into three 8-hour or two 12-hour shifts, depending on the number of RMs available and the situation. If people must walk through a hazardous radiation area to get to food, restrooms, or water, additional RMs may be needed.

1-3. Definitions of Special Terms. A list of special terms and their definitions is given in the glossary in Appendix A.

1-4. What is Nuclear Radiation? In the early 1900s, scientists discovered that certain materials eject three different kinds of energetic rays, which they named alpha, beta, and gamma rays. These rays, or radiation, can pass through certain thicknesses of air, liquids, and solids much like streams of tiny bullets, but at speeds many thousands of times faster than the fastest rifle bullet. The rays cannot be seen, heard, felt, smelled, or tasted.

Later, it was discovered that alpha and beta rays are very tiny particles with an electric charge that move more slowly and are less penetrating than gamma rays (which travel at the speed of light). Still later, it was found that gamma rays are packets of pure energy, called photons, that contain neither electrical charge nor matter. Because of this property of gamma radiation, it is more penetrating than alpha or beta radiation. Visible light and x rays are also composed of photons, except that the photons of light have much less energy than the photons of gamma rays or x rays. Alpha, beta, and gamma radiation originate from the nucleus, or central part, of a radioactive atom. This radiation is, therefore, called nuclear radiation.

All three kinds of radiation, alpha, beta, and gamma, are emitted from radioactive fallout particles produced by the explosion of nuclear weapons. Although alpha and beta radiation can be dangerous under certain conditions, the greatest threat to human life from fallout, and the most difficult threat to protect against, arises from gamma radiation.

1-5. How Radioactive Fallout is Produced. When a nuclear weapon explodes near the ground, it makes a big pit or crater. Tons of earth in the crater are instantly changed from solids into hot gas and fine dust by the tremendous heat and pressure from the bomb explosion. This hot gas and dust, together with vaporized materials, form a giant fireball that rises rapidly in the air to high altitudes. It becomes the top part of the familiar mushroom cloud of a nuclear explosion (Figure 1-1).

Figure 1-1. The mushroom cloud of a nuclear explosion.

Much dust and earth are sucked up with the fireball. Some of this dust and heavier particles make up the stem of the mushroom cloud. The top of the "mushroom" spreads out, cools, and forms a cloud of fine particles of earth and bomb materials. This dust cloud is carried for miles by the wind and drifts down to the earth as fallout. The dust in the stem and in the mushroom cloud becomes radioactive mostly from radioactive materials created in the nuclear explosion that become stuck to part of the dust particles. The air around the particles does not become radioactive, and neither do the surface materials on which they settle. The heavier, large particles settle closer to the explosion than the small particles, which can be carried several hundred miles by the wind. Most of the fallout will come to the ground within 24 hours. Very small particles come down very slowly and may be spread over large areas of the earth's surface, over periods of many days, even weeks. This delayed fallout is sometimes called "worldwide" fallout, although most of the fallout comes down in the hemisphere in which it is produced

(Northern or Southern). Fallout that arrives within the first day or two after the explosion poses a much greater threat to human life than delayed fallout.

1-6. How Nuclear Radiation Harms Our Bodies.

a. Alpha radiation is stopped by the outer skin layers and isn't harmful unless fallout particles are inhaled or swallowed. In this case, the alpha radiation may cause serious damage to the tissues inside the lungs or digestive tract. However, it is unlikely that anyone will breathe or swallow enough particles to become a casualty from alpha radiation during the emergency. The fallout particles are too large to pass through the respiratory tracts without being filtered or trapped, and it is unlikely that anyone will swallow large quantities of fallout particles except under bizarre circumstances. We do not need to be concerned here about alpha radiation from fallout.

b. Beta radiation is much more penetrating than alpha radiation and may cause skin burns if a lot of fallout particles less than a few days old stay on the skin for a few hours. It may also be a greater hazard than alpha radiation if fallout particles are accidentally eaten or inhaled. If fallout particles are accidentally swallowed or inhaled, some of the radioactive atoms will find their way into the bones and organs of the body, where the alpha and beta radiation may possibly cause cancer years later. Again, it is unlikely that anyone will breathe or swallow enough fallout particles to become a casualty from beta radiation during an emergency, for the same reasons as given for alpha radiation.

c. Gamma radiation is the most dangerous of the three kinds of fallout radiation, because it can penetrate the entire body and cause cell damage to all parts, to the organs, blood, and bones. If enough cells in your body are damaged by gamma radiation, you will feel sick after a while. Higher levels of exposure will cause death. Even if you are exposed to enough radiation to make you sick or possibly kill you later on, you may not feel anything while the radiation is causing damage. The reason you don't feel anything is because the nerve cells are not directly stimulated by nuclear radiation as they are by pressure and temperature.

The hazard from nuclear radiation is much reduced within a few days after fallout has arrived because all radioactivity in fallout from nuclear weapons decays by natural processes. The rate of radioactive decay is most rapid during the first few days, and gradually slows as time goes on.

1-7. How We Measure Quantities of Nuclear Radiation. We cannot weigh nuclear radiation or collect it in a box, just as we cannot weigh or collect sunshine in a box. We must measure these things by the effects they cause. Unlike the part of sunshine that we can see, invisible nuclear radiation produces an electrical effect called ionization in the materials it passes through. This ionization can be measured by special instruments.

The roentgen (abbreviated R) is a unit of measurement for exposure to gamma and x-ray radiation. This unit is named after Professor Wilhelm Roentgen, who was the discoverer of x rays in 1895. The harmful effects of nuclear radiation are related to the quantity of radiation to which a person is exposed. The quantity of radiation exposure will be given in units of roentgens. We use two kinds of instruments to measure nuclear radiation. One measures the total accumulated exposure to the radiation, and the other measures the rate of exposure, or how quickly radiation exposure is accumulated.

A DOSIMETER is a radiation detection instrument which gives its readings directly in units of roentgens. These instruments are called dosimeters because they measure the total "dose" or accumulated amount of radiation to which they are exposed.

Another kind of instrument, the SURVEY METER, will measure the rate of exposure, in units of roentgens per hour (abbreviated R/hr). These instruments are called survey meters because they can be used to look over, or survey, an area to find out what the radiation levels are and find the spots where the nuclear radiation intensity is the highest or the lowest.

1-8. How Much Nuclear Radiation is Harmful?

a. Natural Background Levels. Low levels of nuclear radiation are a natural part of our surroundings. Radioactive elements in our own flesh and blood give off nuclear radiation, as they do in the foods we eat, the buildings we live in, and some of the water we drink. Nuclear radiation also comes from the sky and is called cosmic radiation. Nuclear radiation is part of all of our lives and has been present since the earth was formed.

In the United States, the exposure per person to natural nuclear radiation during a whole year is seldom more than two-tenths of a roentgen. These background levels of nuclear radiation are too low to be measured by the radiological instruments provided for shelters. Levels of nuclear radiation from fallout will be thousands of times higher and will be measured in roentgens per hour (R/hr) instead of roentgens per year.

b. Symptoms of Radiation Injury. Although nuclear radiation from the natural background damages some cells in our bodies and destroys others, we do not notice this damage. Billions of cells in our bodies die natural deaths every hour and are replaced by normal growth and repair processes. We feel no injury or sickness from exposure to nuclear radiation at the levels which exist in our natural surroundings.

But if our bodies are exposed to gamma radiation from fallout which is many thousands of times higher than the levels of natural background nuclear radiation, there will be so many cells damaged or destroyed that some of us may become sick, and some may even die. Some or all of the symptoms of injury may appear within the first three days after exposure. These symptoms include nausea, vomiting, diarrhea, fever, irritability, a lack of energy, and a feeling of being tired. The symptoms may disappear and then come back after a week to three weeks later, sometimes with diarrhea, sore throat, loss of hair, and a tendency to bleed easily. The greater the exposure, the earlier the symptoms will appear. They will also be more severe and last longer. Chances of illness from infections are greater among those who are exposed to more than about 200 R, because the high radiation exposure damages the immune system in our bodies that helps fight diseases. In small children, the symptoms of radiation injury will appear at lower exposures than for adults. Even though a person may be severely affected by high radiation exposures, the person does not become radioactive from such exposure, and will not be a radiation hazard to anyone else.

Beta burns will result if a lot (enough to make you feel dirty or grimy) of fallout particles less than a few days old stay on the skin for several hours. Early symptoms of such skin contamination include itching and burning sensations. These may soon disappear. Darkened or raised skin areas or sores may appear within one or two weeks. After two weeks or more, there may be a temporary loss of hair (it will return in about six months). The greater the exposure, the earlier the symptoms will appear. Beta burns will not be a problem if fallout particles are brushed or washed off promptly. Wearing clothing such as gloves, hats, scarves, face-masks, and long-sleeved garments will help to prevent fallout particles from collecting on the skin. Within a few days after fallout has arrived, its radioactivity will have decayed so much that beta radiation will not be a hazard under most circumstances. It may be a problem in the first few weeks if a person must lie or crawl on the ground, as may be necessary in rescue operations, and the skin is covered with dust which is not removed for many hours.

We are concerned mostly about radiation injury from gamma radiation from fallout particles on the ground, buildings, trees, and shrubs around us. This radiation is called external radiation because it comes from particles which are outside the body.

c. Effects and Levels of Sickness from Brief Exposure. When people are exposed to gamma radiation from fallout, their entire bodies are exposed, including arms, legs, head, and trunk. This kind of exposure, called whole-body exposure, differs from medical exposures in which radiation may be concentrated on one small part of the body. A whole-body brief exposure to 50-200 R of gamma radiation may result in radiation sickness, but if only a part of the body such as the hand or foot is exposed to 50-200 R of gamma radiation, as in medical treatment, there will be no radiation sickness.

The human body has ways of repairing damage done to it. Because of these repair mechanisms, a whole-body radiation exposure of 600 R spread out uniformly over a period of 20 years would not cause any radiation sickness. But if this exposure were received over a brief period of a week or less, it would probably result in death.

Some people may become very sick within a few weeks after being exposed for a brief time (a week or less) to a certain amount of gamma radiation from fallout. Others may be exposed to the same dose and not feel any serious effects. If the exposure is less than 50 R, the injury from radiation should not produce symptoms in anyone. Some people irradiated in this dose range might experience loss of appetite and nausea, but this could also be the result of anxiety and fear.

Doctors have described five levels of sickness which occur after brief whole-body exposure to 50 R or more of gamma radiation from fallout. These levels are described here and summarized in Table 1-1. Additional effects are described in the books listed in the Bibliography, Appendix C.

Table 1-1. Levels of sickness and probable conditions of most people after brief whole-body exposure to gamma radiation

Exposure Range (roentgens)	Response	Probable condition of majority during emergency		Probable death rate during emergency	Comments
		Medical care required	Able to work		
0-50R	No symptoms	No	Yes	0	
50-200R	Radiation sickness, Level I	No	Yes	Less than 5 percent	Deaths will occur in 60 or more days
200-450R	Radiation sickness, Level II	Yes	No[a]	Less than 50 percent	Deaths will occur within 30-60 days
450-600R	Radiation sickness, Level III	Yes	No[a]	More than 50 percent	Deaths will occur in about one month
More than 600R	Radiation sickness, Levels IV & V	Yes	No	100 percent	Deaths will occur in two weeks or less

[a] Except during illness-free latent period.

(1) Level I, 50-200 R Exposure. Less than half of the people exposed to this much radiation experience nausea and vomiting within 24 hours. Afterwards some people might tire easily, but otherwise there are no further symptoms. Less than five percent (one out of 20) need medical care. Any deaths that occur after radiation exposure are probably due to additional medical problems (complications) a person might have at the same time, such as infections and diseases, injuries from blast, or burns from the nuclear explosion.

(2) Level II, 200-450 R Exposure. More than half of the people exposed to 200-450 R experience nausea and vomiting and are ill for a few days. This illness is followed by a period of one to three weeks when there are few if any symptoms (a latent period). At the end of this latent period more than half of those exposed experience loss of hair. A moderately severe illness develops which is often characterized by sore throat. Radiation damage to the blood-forming organs results in a loss of white blood cells, increasing the chance of illness from infections. Most of the people in this group need medical care, but more than half will survive without treatment. The chances of living are better for those with smaller doses and those who get medical care. More than half are sick the first few days, but less than half die.

(3) Level III, 450-600 R Exposure. Most of the people exposed to 450-600 R experience severe nausea and vomiting and are very ill for several days. The latent period is shortened to one or two weeks. The main episode of illness which follows is characterized by much bleeding from the mouth, throat, and skin, as well as loss of hair. Infections such as sore throat, pneumonia, and enteritis (inflammation of the small intestine) are common. People in this group need intensive medical care and hospitalization to survive. Fewer than half will survive in spite of the best care, the chances of survival being poorest for those who received the largest exposures.

(4) Level IV, 600 to over 1000 R Exposure. This level produces an accelerated version of the illness described for Level III. All the people in this group begin to experience severe nausea and vomiting. Without medication, this condition can continue for several days or until death. Death can happen in less than two weeks, without the appearance of bleeding or loss of hair. It is unlikely, even with extensive medical care, that many can survive.

(5) Level V, Several Thousand Roentgens Exposure. Symptoms of rapidly progressing shock come on almost as soon as the dose has been received. Death occurs in a period from a few hours to a few days. It is highly unlikely that exposures of this magnitude will be experienced in fallout shelters.

d. Long-Term Effects. In addition to early sickness, exposure to nuclear radiation has some effects which may not show up for months or years. In a nuclear war, our first concerns will be with survival from the early effects. If the levels of nuclear radiation are low enough so that early radiation sickness is not a serious factor, then we become concerned with avoiding long-term effects. After a period of months to years has passed following an exposure to nuclear radiation levels many times higher than background levels, some of the people (less than a few percent) may develop various kinds of cancers. The probability of developing such late effects should not be used as a principal determining factor in decision-making during a war emergency, but such effects can and should be minimized by keeping controllable exposures as low as practicable.

There may also be effects on babies exposed while in the womb and genetic effects in children whose parents (one or both) were exposed to high levels of radiation. Pregnant women who are exposed to enough radiation to cause symptoms of early radiation sickness (over 50 R), as described above, may have a miscarriage. There may be some developmental defects in the few babies born to the heavily exposed mothers.

Additional information on long-term effects may be found in the books listed in the Bibliography, Appendix C.

e. Contamination of Food and Water. Food and water that have been exposed to nuclear radiation but are not contaminated by fallout particles are not harmed and are fit for human consumption, unless spoiled in some other manner. If food containers, fruits, vegetables, and grains become contaminated by the presence of radioactive fallout particles, they need not be thrown away. If the particles can be removed by washing, scrubbing, brushing, or peeling, the food is safe for consumption.

Water in covered containers and from underground sources will be safe. Water into which fallout particles have fallen may become unsafe to drink for a while, because radioactive iodine dissolves in water. Water that is collected from rooftops or other flat areas into cisterns, tanks, or other reservoirs, may have much higher concentrations of radioiodine than other sources of water if there is a rainfall shortly after fallout has arrived. Rivers and streams that are fed mostly by water from the surface rather than from underground springs may also become contaminated by radioiodine if there is a rainfall in the first few days after fallout arrives. Water in large, deep lakes, reservoirs, and rivers will probably be safe to drink (although it could still be unsafe due to other pollutants) within several hours or days after fallout has arrived, because of dilution of the radioiodine into large volumes of water.

The radioiodine problem will almost completely disappear in any water in a few weeks due to natural radioactive decay. The quantity of the radioactive iodine of greatest concern will become half as much every eight days (the half-life is eight days).

Radioisotopes that have dissolved in water cannot be removed by boiling or settling. The water can be purified by special filtering or chemical processes, one method being the filtration of water through several inches of soil or clay (not sand). Water filtered through soil must be disinfected either by boiling or by adding chemicals such as chlorine or household iodine.

Even low levels of radioiodine in water, undetectable by the radiological instruments described in this handbook, may cause special medical problems in some people. Radioiodines may be concentrated and stored in the thyroid gland, resulting in possible radiation damage to the thyroid in some people and, after several months or years, thyroid tumors, hypothyroidism, or thyroid cancer. These symptoms, except radiation damage, may be treated medically. Infants and fetuses are more susceptible to this threat than older children and adults. In any case, NO ONE SHOULD BE DENIED WATER BECAUSE OF POSSIBLE FALLOUT CONTAMINATION.

The concentration of radioiodine in the thyroid gland can be blocked by taking pills or drops of potassium iodide (KI). If the gland is saturated with nonradioactive iodine, it will not take up radioactive iodine that enters the body by any means. Taking potassium iodide does not provide protection against gamma radiation from outside the body.

The Food and Drug Administration has declared potassium iodide for use in radiation emergencies to be a nonprescription drug. Potassium iodide pills may be obtained at many drugstores. Some states are stockpiling potassium iodide pills to be used in the event of a radiation emergency involving a nuclear reactor.

For children and adults, the recommended dose of potassium iodide is 130 milligrams taken by mouth each day for 14 days, unless it becomes known before the 14 days are up that the drinking water is not contaminated or a source of uncontaminated water is found. If the same drinking water must be used all the time without knowing whether it is contaminated, the potassium iodide dose may be reduced after 14 days to half a pill (or 65 milligrams) taken by mouth each day by children and adults. Children under one year of age may be given half the dose taken by adults, although the adult dose would be safe. Taking more than one 130-mg tablet per day will not improve the blockage of radioiodine. For best results, the dosages should be started preferably before drinking water suspected of being contaminated but not later than three to four hours after drinking such water.

Side effects from this dosage are expected to be very rare. Persons with KNOWN ALLERGIES to iodine should NOT TAKE this medication.

1-9. <u>How You Can Shield Yourself from Gamma Radiation</u>. You can protect yourself from bullets by surrounding yourself with armor plate. In a similar way (but not exactly!), you can shield yourself from gamma radiation. Anything between you and the source of gamma radiation will cut down the number of rays that reach you. The heavier or more massive the barrier between you and the source, the more the radiation is cut down.

A wall of concrete will give better protection than a wall of earth of the same thickness, because the concrete wall is heavier. Concrete has a greater density than earth. But if the concrete wall is thinner than the earth wall (but the same height and width) so that the overall weight is the same, each wall will give the same protection. It's not the thickness but the total weight of material between you and the fallout that is important.

A wall of a certain thickness will stop a bullet of a certain size and speed without any doubt. For gamma rays, a wall only cuts down the chances of the gamma "bullets" getting through. A wall of concrete eight inches thick will cut down the gamma radiation from fallout by about a factor of 10. Other materials will reduce the radiation more or less effectively, depending on whether their specific gravity or density is greater or less than that of concrete.

To give you an idea of which common materials might be useful for shielding, a list of materials is given in Table 1-2, showing their densities compared with concrete. Note that earth is almost as good as concrete and is usually available and inexpensive. Water and gypsum wallboard are better shielding materials than wood and newspapers. Lead is the most dense shielding material of those listed. Lead bricks and sheets are often used as shielding material where little space is available, but are too expensive for general use in shelters.

Table 1-2. A list of materials to give you an idea how good they would be as barriers against gamma radiation from fallout

(Materials with the highest densities require less thickness to cut down the gamma penetration by a given amount)

Material	Density relative to concrete[a]	Thickness required relative to concrete
Aluminum	1.2	0.8
Brick, common clay	0.7	1.4
Concrete	1.0	1.0
Earth (well-packed moist humus, dry clay)	0.7	1.4
Firebrick (used in fireplaces)	0.9	1.1
Glass	1.1	0.9
Hardwood (maple or oak)	0.3	3.3
Human Body	0.4	2.5
Lead	4.9	0.2
Magazines, slick	0.4	2.5
Newspaper (flat), books, pulp magazines	0.3	3.0
Plywood (dry)	0.2	5.0
Steel	3.4	0.3
Wallboard, gypsum	0.4	2.7
Water	0.4	2.3

[a] Concrete of density 2.3 g/cm^3 (144 lb/ft^3).

When fallout particles are all around you on the ground outside the shelter, you will need a barrier all around you to shield yourself from gamma radiation. You will also need a barrier ABOVE you, even though the fallout particles may have already settled to earth. An overhead barrier is needed because gamma radiation is scattered by air, somewhat like auto headlights are scattered by fog. The scattered gamma radiation can reach you from above. This scattered radiation from above is called "skyshine" and is not as penetrating as the radiation coming in a straight line directly from fallout. Fallout may settle on roofs or hillsides above your shelters, and the direct radiation from this fallout will add to the overhead radiation from scattered gamma rays. If you are in a shelter which is below ground, you will need to be concerned mostly with overhead radiation.

Gamma radiation is also scattered around the corners of tunnels and corridors by the air and by the material in the walls. The intensity of gamma radiation scattered around corners is much less than that of the direct radiation.

Because of the penetrating and scattering nature of gamma radiation, the unevenness of fallout, and the different thicknesses of materials between you and the fallout at different places in the shelter, you will need to use the radiation survey meter to find places where the radiation levels are lowest in the shelter. You may improve the shelter and make the radiation still lower by using available materials to build barriers between you and the strongest sources of radiation. The survey meter will tell you how well you have succeeded.

1-10. How Fallout Radioactivity Arrives and Decays. As fallout settles on the shelter and its surroundings, the needle on the survey meter may climb steadily for some time. On the other hand, if you are well protected or if the fallout is not heavy in your location, you may see little or no indication on the survey meter. The fallout cloud or clouds may take as little as 15 minutes or as long as several hours to arrive and begin to deposit fallout in your area.

a. Fallout from One Weapon. If the fallout comes from only one relatively small nuclear weapon exploded on or near the ground less than 20 miles upwind, the fallout may start to fall on the shelter in less than an hour after the explosion. After the fallout begins, it may keep on coming down for an hour or so. If you are outside (where you should not be unless you are on your way to shelter or there is an extreme emergency) you may be able to see some very fine particles coming down. You may also notice a darkness in the sky and feel gritty particles strike your face. After several minutes, a buildup of a thin layer of fallout dust may be noticeable on the tops of cars and on window ledges. If the fallout is visible, the radiation levels are hazardous.

b. Fallout from Many Weapons. If the fallout comes from many large weapons exploded on or near the ground 100-200 miles upwind, the fallout may not start to fall on your shelter for many hours. The time it takes for the fallout clouds to arrive at your location depends on how far upwind the explosions were and on how fast the winds carry the clouds. After the fallout begins, it may keep on coming down for several hours. Larger particles from the explosions will fall to the ground faster than the smaller particles. The clouds will contain mostly very small particles or fine dust by the time they arrive at the shelter, if it is well downwind of the explosion. This dust may be too fine to feel or see, although a darkening of the sky may be noticeable. The buildup of dust on surfaces will be gradual and won't be obvious. The nuclear radiation exposures from this almost invisible fallout may be just as great as, or greater than, the radiation exposures from the more visible fallout from explosions that are closer.

There will be a cloud of fallout particles formed by each of the ground explosions. Some of the clouds may merge. As each cloud with its trail of fallout passes over your shelter, the needle on your radiation survey meter may climb to higher levels.

c. After Fallout Stops Coming Down. After a fallout cloud passes by and when almost all the fallout particles from that cloud have reached the ground in your area, the survey meter needle will slowly begin to fall as the radioactivity from fallout decays and fades away by natural processes. The radioactive materials produced by the nuclear bomb explosion are unstable. These materials change (or decay) into a stable condition by shooting out nuclear radiation. Some materials decay into their stable form faster than others. Those that change fast are very busy producing intense nuclear radiation in the first few moments after a nuclear explosion. Those that decay more slowly may be responsible for measurable nuclear radiation years after the explosion.

d. The Seven-Ten Rule. Because many materials in the fallout cloud decay quickly, the nuclear radiation from a given quantity of fallout is most intense in the first moments after detonation and its intensity rapidly falls to lower levels. This behavior can be approximately described by a rule of thumb called the seven-ten rule. This rule APPLIES ONLY TO FALLOUT OF THE SAME AGE. If the fallout at a location is a mixture resulting from detonations that took placc at different times, the seven-ten rule does not apply.

THE SEVEN-TEN RULE states that the measured radiation intensity from a given quantity of fallout particles will decay to (1) ONE-TENTH AS MUCH when the fallout becomes SEVEN TIMES OLDER than the age at the time of measurement, (2) ONE-HUNDREDTH (1/10 x 1/10) AS MUCH when the fallout becomes FORTY-NINE TIMES (7 x 7) OLDER than the age at the time of measurement, and so forth. The unit of time can be seconds, minutes, hours, half-days, days or whatever period of time is appropriate for the situation. The seven-ten rule is illustrated in Figure 1-2.

Suppose that fallout begins to arrive at a certain location two hours after the explosion, and the fallout keeps on coming down for three more hours before it stops. The seven-ten rule cannot be used with survey-meter readings taken while fallout is arriving, because the quantity of fallout is increasing. At five hours after the explosion, when the fallout at this particular location has clearly stopped, the survey-meter reading can be used with the seven-ten rule to predict the future radiation intensity at the location, if no additional fallout arrives, and weather doesn't change the fallout pattern. In this case, the age of the fallout (or the time "T" after detonation, according to Figure 1-2) is five hours.

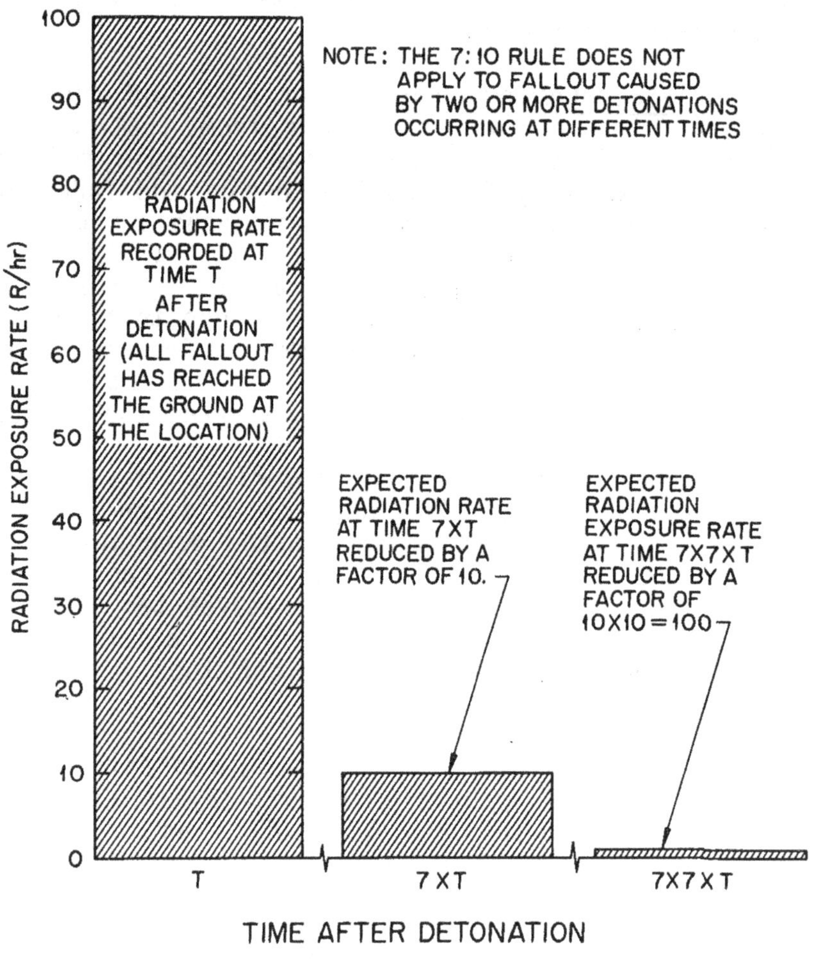

Figure 1-2. Illustration of the seven-ten rule. The time "T" after detonation is the age of the fallout.

The seven-ten rule states that the radiation intensity at this location at 7 x 5 hours, or 35 hours after the explosion, will be reduced by a factor of 10 to 1/10th, or 10 percent, of the reading obtained at five hours. Furthermore, the seven-ten rule states that for two multiples of seven, or at 7 x 7 x 5 hours, or 7 x 35 = 245 hours (10 days and 5 hours) after the explosion, the radiation intensity at this location will be reduced by two factors of ten to 1/100th (1/10 x 1/10), or one percent, of the reading obtained at five hours after the explosion.

If another reading is taken at this location when the fallout is nine hours old, the seven-ten rule states that the intensity at 7 x 9 = 63 hours after the explosion will be reduced by a factor of 10 to 1/10th the reading at nine hours, and the intensity at 7 x 7 x 9 = 441 hours (18 days, 9 hours) will be reduced by a factor of 100.

If there are several nuclear ground bursts detonated at different times upwind or if there is a heavy rain during or after the fallout, the seven-ten rule doesn't apply. Other rules have been developed to forecast upper limits of radiation exposure, as described in paragraph 4-5a.

e. Radioactive Decay. Decay of the radiation intensity from radioactive fallout particles takes place in the cloud as it is carried by winds toward you. The radiation intensity will also be decreasing because the cloud spreads out as it moves along, and the heavier particles will be dropping out, so the number of fallout particles per cubic inch of air will be decreasing as time goes on. Radioactive materials in the clouds from distant explosions will have more time to decay and spread out while they are on their way. Many of the materials that decay quickly will have decayed to undetectable levels before reaching you. For this reason, the radiation intensity from fallout on the ground from distant explosions will decrease more slowly when it reaches the ground than the radiation intensity of fallout from closer explosions.

f. Rainout. If the air is humid, the nuclear explosion may start a local rain. If it is already raining or if the explosion starts a rain shower, much of the radioactive material will come quickly to the ground as "rainout." When rainout occurs soon after an explosion, the fallout cloud has not had a chance to spread out as it does when carried a long way by the wind, and it has not had as much time to decay.

If the rainfall producing rainout is light, local radiation intensities may be much higher than when produced by dry fallout. If the rainfall is heavy, the radioactive material may be washed into gutters, ditches, and storm sewers. From there, it may flow into streams and rivers. Radioactive materials, like dirt particles, can collect in unpredictable locations under these circumstances. The radiation survey meter will be needed to help detect, and avoid remaining in, such locations.

CHAPTER 2

INSTRUMENTS FOR DETECTING NUCLEAR RADIATION

2-1. What Is Needed. If radioactive fallout settles on a shelter and its surroundings, people in the shelter will want to know where to go and what to do for the best protection. People in your shelter will want to know whether they are going to get sick and possibly die from radiation exposure. After the worst radiation has faded away, they will want to know the risks of going outside, how long they can stay outside, and which locations will result in the least radiation exposure. To answer these questions, you will need special instruments.

The levels of radiation from fallout from nuclear weapons can be much higher than those encountered in peacetime conditions. The radiation instruments developed for use by operators of nuclear reactors, by radiation therapists in hospitals, or by crewmen of nuclear submarines and ships are generally not suitable for the needs of people caught in the radioactive fallout of a nuclear war. These commercial instruments for peacetime purposes usually do not have the higher ranges which may be needed for wartime use.

To meet the special needs of people who may face radiation hazards from radioactive fallout that may result if this country is attacked by nuclear weapons, the U.S. Government has developed two kinds of radiological instruments. The SURVEY METER is designed to help you find the places of lowest radiation intensity and to indicate where you should not go because of high radiation levels. The DOSIMETER is designed to help you estimate the total amount of radiation to which your body has been exposed. Without a dosimeter it would be difficult to estimate exposure to radiation if you need to move around in places where there are different radiation intensities or if the radiation intensity rises and falls irregularly due to fallout from passing fallout-laden clouds.

2-2. What If There Are No Instruments? If people have assembled in a shelter in a nuclear war emergency and there are no radiation detection instruments, try to obtain these instruments from your local government before fallout arrives (see STEP 2a of Checklist "A"). If no radiological instruments can be obtained, try to find the location in your shelter that you think will provide the best protection from nuclear radiation (see paragraph 4-2b, "Checking out the Shelter"). Listen to your local radio station, particularly one which is tied in with the Emergency Broadcast System (EBS), for news of approaching fallout.

If there are no radiological instruments, then you will need communications with those who have them. You will need information from others, from your local Emergency Operating Center, if possible, or from EBS. If you have no radiological instruments, communications may provide the only warning of the arrival of a radiation hazard. Remember, although the particles may be seen, heard, and felt under some conditions while they are coming down, the fallout radiation itself is invisible and silent and cannot be felt.

2-3. The Survey Meter. A survey meter is illustrated in Figure 2-1. The gamma-radiation level (exposure rate) is shown by the position of the needle on the instrument dial. When the needle points to a number on the dial, that number, when multiplied by the range-selector number (described in paragraph 3-2d(2)), will tell you the level of gamma radiation from fallout in units of roentgens per hour (R/hr) at the location of the instrument.

Figure 2-1. A survey meter.

Gamma rays pass through the metal case of this instrument and also through the walls of a metal can (called the ionization chamber) inside the case. The ionization chamber is sealed to keep out moisture and dust and to maintain a constant pressure. Some of the gamma rays produce charged particles inside the ionization chamber, and these charged particles are collected to make a tiny electric current. This electric current is amplified by electronic circuits in the survey meter to make a much stronger current, which moves the needle. If the survey meter is moved to a location where there is negligible gamma radiation, the needle will return to the zero position.

Radiation levels from fallout in a nuclear war can be very high. The highest reading on the survey meter is 500 R/hr. Because of the large range of radiation levels which you might encounter, from low to very high, this instrument was designed with a range-selector switch. If this instrument had no range-selector switch, a low but still hazardous radiation exposure rate of 5 R/hr would cause an almost undetectable needle movement. This situation would compare with trying to read the speed of a car going one mile per hour on a speedometer that reads 100 mph full scale. By switching the range-selector switch to a different position, the maximum range of the needle can be changed from 500 R/hr to 5 R/hr. The radiation exposure rate of 5 R/hr would then cause the needle to swing all the way through its full range of movement to the high end of the scale. With another position of the range-selector switch, readings as low as 0.05 R/hr can be read accurately.

Instructions on how to get the survey meter ready for operation and how to use it are given in paragraph 3-2.

2-4. The Dosimeter. A dosimeter is shown in Figure 2-2 with a ballpoint pen for comparison of size. The dosimeter has a clip so it can be attached to clothing worn on the body. It is usually worn in a breast pocket. If a person's clothing has no breast pockets, the dosimeter can be clipped to the collar, neckline, or belt. In some situations, dosimeters may be mounted on walls, posts, or furniture or hung by string.

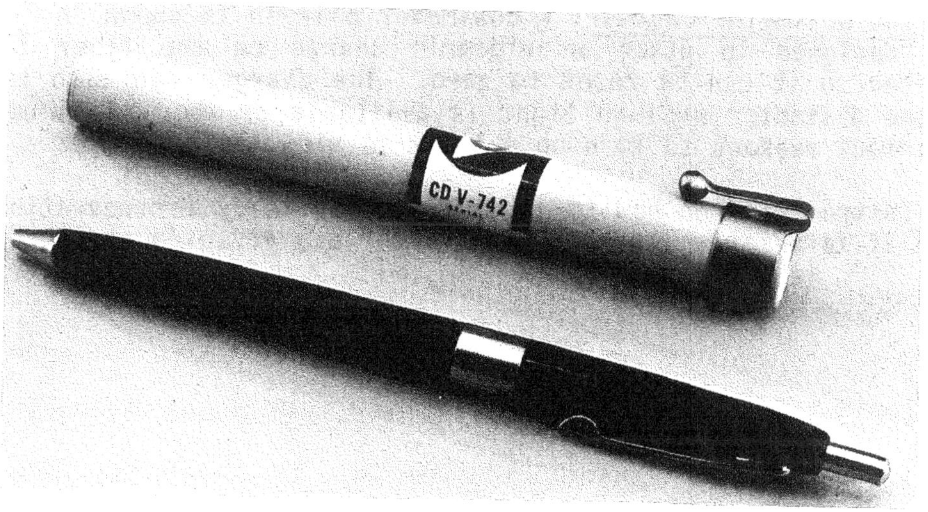

Figure 2-2. A dosimeter shown with a ballpoint pen for comparison of size.

The dosimeter shows the total or accumulated amount of gamma radiation to which it has been exposed starting from the time of recharging (or zeroing) the instrument. This gamma exposure is read by holding the instrument so that it is pointed toward a bright light and looking through one end, the end with the clip on it. The gamma exposure is shown by the position of a hairline along a scale of numbers marked "ROENTGENS." The scale has numbers that begin with zero at the left side and usually end with 200 at the right side.

The dosimeter is constructed to be reliable and rugged. The only moving part is the hairline or fiber seen through the eyepiece. Its design is based on the principle that a charge of electricity is reduced when there are charged particles around, and charged particles are produced by gamma radiation. A special instrument is used to place a charge of electricity inside the dosimeter. This charge is just like the static electricity that builds up on a person who is walking along a carpet on a dry winter day. The position of the fiber depends on how much static electric charge is on it. When gamma rays interact with the walls of the dosimeter and enter the chamber in which the fiber is sealed, charged particles are produced. These particles reduce the charge on the fiber, and the fiber moves to a different position. The position of the fiber as it is seen on the scale then indicates the total amount of gamma radiation to which it has been exposed since it was charged.

Instructions on how to get the dosimeter ready for operation and how to use it are given in paragraph 3-4.

2-5. The Dosimeter Charger. A dosimeter charger is shown in Figure 2-3. It is designed to place an electric charge on the fiber inside the dosimeter so it can be reset to zero. The charger can also be used to read the dosimeter when no light is available or when it is undesirable for various reasons to turn on a light to view the dosimeter scale.

Instructions on how to get the charger ready for operation and how to use it to reset and read the dosimeter are given in paragraph 3-3.

Figure 2-3. Dosimeter charger.

CHAPTER 3

HOW TO GET YOUR RADIOLOGICAL INSTRUMENTS READY FOR OPERATION

3-1. Before You Begin. Radiological instruments could save your life, so treat them with respect! Don't drop them, don't spill liquids on them or immerse them, and don't let children play with them. If you have never used these instruments READ ALL OF THE FOLLOWING INSTRUCTIONS BEFORE YOU TRY TO OPERATE THE INSTRUMENTS.

One person should be designated to be responsible for the care and operation of each survey meter in the shelter. Other persons, or perhaps the same persons, depending on the number of people in the shelter, should be designated to be responsible for the care and use of dosimeters and dosimeter chargers.

3-2. Preparation for Using the Survey Meter.

a. Preliminary. What the survey meter measures and how it works are briefly described in paragraph 2-3. The survey meter (Figure 2-1, page 2-2) has two controls: the range-selector switch underneath the handle and the zero control on the corner. A carrying strap will be appreciated in a fallout situation when you may need to use your hands to do something else and you don't want to put down the instrument.

b. Installing the Battery in the Survey Meter. The survey meter is powered by a single D-cell flashlight battery. The battery is installed as follows:

(1) Turn the range-selector switch (the switch underneath the handle) to the "OFF" position.

(2) Open the case by unfastening the case clips at each end.

(3) Use the handle to lift the top part of the survey meter out of the bottom part of the case. The top may be laid on a flat surface or held in the hand by the handle while installing the battery.

Don't let dust, sand, or moisture get in the case. If fallout particles get inside the case, you will get a false reading! Also, don't let anyone touch the circuit board or other interior parts. Grease or sweat on the electronic components may cause a malfunction.

There may be a packet inside the survey meter which may or may not be labelled "desiccant." Leave the packet inside the case. Don't get it wet! It will help keep the inside dry. Dryness is necessary to prevent small electric currents from leaking across insulators and to prevent corrosion.

(4) Install the D-cell battery in the rectangular plastic battery holder that is mounted on the inside of the top cover. The inside of the top cover with a battery inserted is shown in Figure 3-1. One end of the floor of the battery holder will be marked with a plus sign (+) and the other end with a negative sign (-). The battery may also be marked with these signs but, if it isn't, the positive end can be identified by the raised center post. Insert the battery in the holder so the plus sign (+) on the battery or the positive electrode is on the end where the plus sign (+) is marked on the floor of the holder. Push the battery in firmly so the metal electrode clips on the ends of the holder snap over the battery ends and the battery is down in the holder as far as it will go.

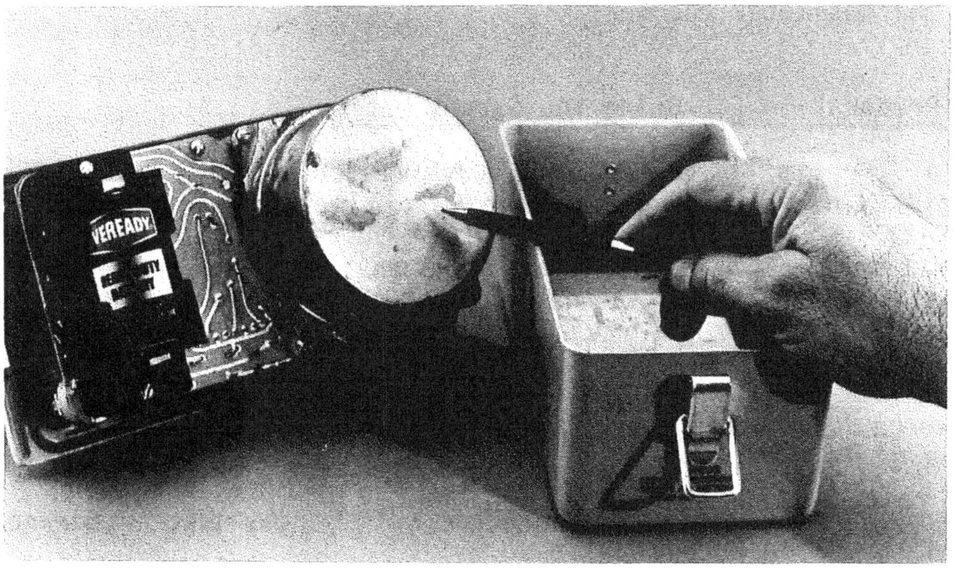

Figure 3-1. A survey meter with the top removed. The top is shown upside down on the left. The pen points to the ionization chamber.

(5) Lower the top part of the survey meter into the bottom part of the case. If there is a small rubber pad glued on one end of the inside floor of the bottom part, turn the case so the pad lines up under the battery.

(6) Fasten the case clips.

c. <u>Checking the Battery and the Instrument (Operational Check)</u>. Every time a battery is inserted in the survey meter, an operational check should be made to make sure the battery has been put in correctly and that it has enough energy to run the survey meter. An operational check should also be made each time before using the survey meter to make sure the meter is operating properly. An operational check is made as follows:

(1) Turn the range-selector switch (the switch underneath the handle) to the "ZERO" position. Wait a full two minutes before doing anything else with the meter. One of the components in the survey meter is a special electronic tube (an electrometer tube), which must be warmed up before it can operate properly.

(2) After waiting two minutes for warmup, rotate the knob marked "ZERO" (the knob on the corner) until the needle on the meter points to "0" (zero). If the needle doesn't move when the ZERO knob is rotated, turn the range-selector switch to "OFF" and remove the battery. Clean the battery contacts and install a new battery, unless the old one is known to be good. If the needle still doesn't move with rotation of the zero control, then the instrument is faulty and should be returned for replacement, if you have the time and opportunity.

(3) After the instrument has been zeroed, turn the range-selector switch to "CIRCUIT CHECK" and hold it there against the spring pressure which will return the switch to "OFF" when the switch is released. While holding the switch in the "CIRCUIT CHECK" position, the needle should climb to the upper part of the meter scale in or near the area marked "CIRCUIT CHECK." A reading of 3 or higher (even though the area marked "CIRCUIT CHECK" begins at 3.5, not 3.0) will tell you three things: (a) the battery was installed properly, (b) the battery has enough energy to run the meter, and (c) the circuits involved in this part of the test are operating properly.

If the needle does not climb up to 3 or higher while the range-selector switch is being held on "CIRCUIT CHECK," remove the battery, clean the battery contacts, and install a new battery, unless the old one is known to be good. Repeat the steps above, including the zero adjustment. If the needle still does not climb up to 3 or higher during the circuit check, the instrument is faulty and should be returned for replacement, if there is a place close enough where you may replace it and get it back before fallout arrives.

(4) After the survey meter has passed the circuit check satisfactorily, rotate the range-selector switch to each of the positions marked "X100," "X10," "X1," and "X0.1." Let the switch rest at each position momentarily, and observe the position of the needle on the meter. If there is no gamma radiation present besides that from normal background radiation, the needle should remain approximately at zero at each position of the switch. If it moves up-scale, it should not move up more than three of the smallest divisions (not above 0.3 reading on the dial) when the range-selector switch points to "X100," "X10," or "X1." When the range-selector switch points to "X0.1," the needle should not move up-scale from zero more than six of the smallest divisions (not above a 0.6 reading on the dial).

This smallest needle movement, called up-scale leakage, will not affect the usefulness of the survey meter in detecting hazardous radiation levels from fallout. If the up-scale leakage is greater than the limits stated above, the amount of up-scale leakage usually can be reduced significantly by leaving the instrument on for one to 16 hours with the range-selector switch in the "ZERO" position. This procedure reconditions the electrometer tube. If excess up-scale leakage still exists after 16 hours of reconditioning, and there is no fallout gamma radiation present, other problems exist. The instrument should be returned for replacement if you have the time and opportunity.

Be sure to turn the range-selector switch to the "OFF" position when the survey meter is not in use.

d. Reading the Survey Meter. After the operational check has been made, the survey meter can be used to measure the gamma radiation exposure rate at the location of the meter, as follows:

(1) Hold the meter steadily in one location at about three feet off the floor (waist height) and about two feet away from your body. The meter is to be held away from your body to reduce the effect of shielding gamma radiation with your body. Turn the range-selector switch clockwise (from "X100" to "X10," then from "X10" to "X1," etc.) until you find the range position that results in the highest reading of the needle on the dial (not over 5). Pause a moment or two at each range position to see how fast the needle climbs.

(2) With the range-selector switch in the "X0.1" position, it will take 10-15 seconds for the needle to stop moving. It will take less time for the needle to reach a steady reading when the range selector switch is at the higher multipliers ("X1," "X10," and "X100"). There are five numbers printed on the dial, starting with "0" on the left and ending with "5" on the right. Between each printed number and the next, there are ten divisions. The dial reading is obtained by writing down the number that appears nearest the needle on its left side, placing a decimal point to the right of that number, and then writing down the number corresponding to the number of the nearest division mark to which the needle points to the right of the printed number. For example, the dial reading in Figure 3-2, page 3-5, is 1.4. In Figure 3-3, page 3-5, the dial reading is 0.4, and in Figures 3-4 and 3-5, page 3-6, the dial readings are 4.1 and 2.5, respectively.

(3) The radiation exposure rate is obtained by multiplying the dial reading by the number following the "X" at the position to which the range-selector switch points. For example, in Figure 3-2, page 3-5, the dial reading is 1.4 and the range-selector switch points to "X100," so the radiation exposure rate is 1.4 x 100, or 140 roentgens per hour (R/hr). In Figure 3-3, page 3-5, the dial reading is 0.4 and the range-selector switch points to "X10," so the radiation exposure rate is 0.4 x 10, or 4 R/hr. Additional examples are shown in Figures 3-4 and 3-5, page 3-6.

(4) When the dial reading is 0.5 or less, the range-selector switch should be switched one position clockwise to get a more accurate reading. In this position, where the switch points to a lower

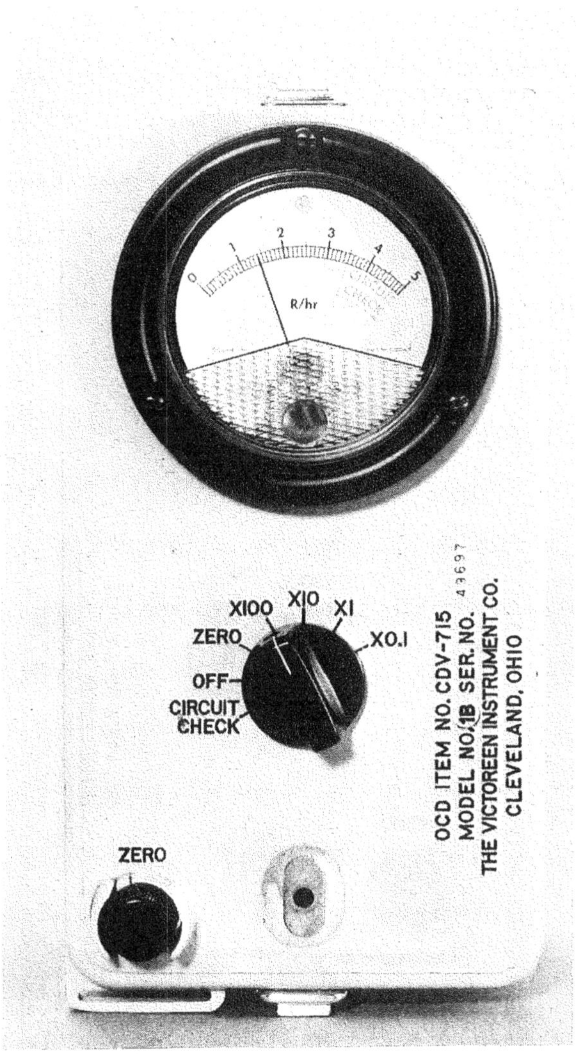

Figure 3-2. The survey meter dial reading is 1.4. This reading is multiplied by 100 to get the radiation exposure rate reading because the range-selector switch points to "X100." The radiation exposure rate is 140 R/hr.

(1.4 x 100 = 140)

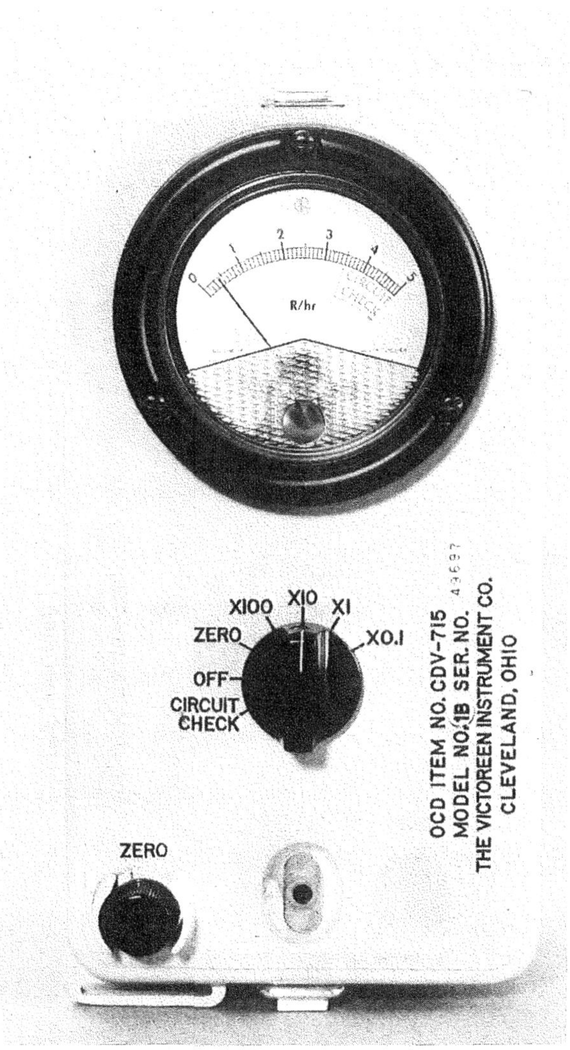

Figure 3-3. The survey meter dial reading is 0.4. This reading is multiplied by 10 to get the radiation exposure rate reading because the range-selector switch points to "X10." The radiation exposure rate is 4 R/hr.

(0.4 x 10 = 4.0)

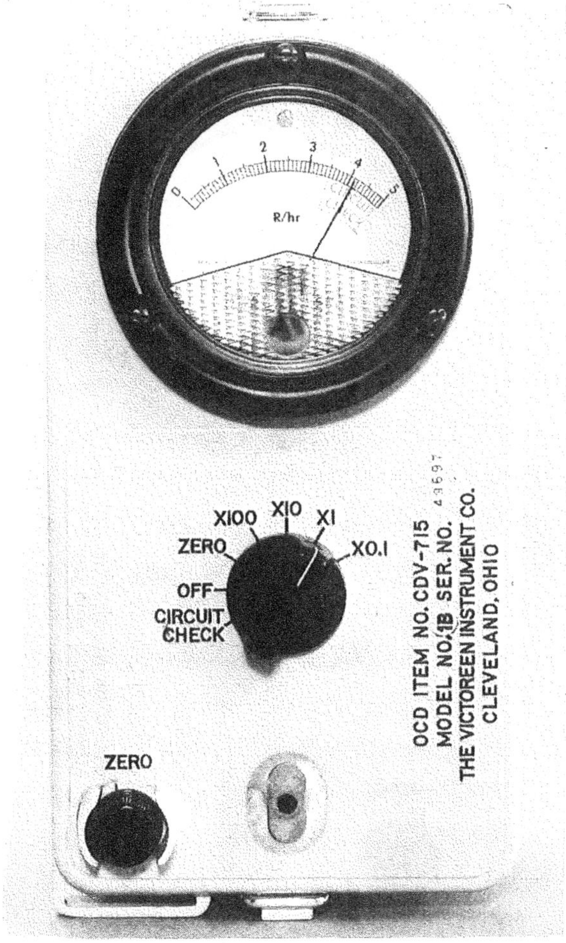

Figure 3-4. The survey meter dial reading is 4.1. This reading is multiplied by 1 to get the radiation exposure rate reading because the range-selector switch points to "X1." The radiation exposure rate is 4.1 R/hr.

(4.1 x 1 = 4.1)

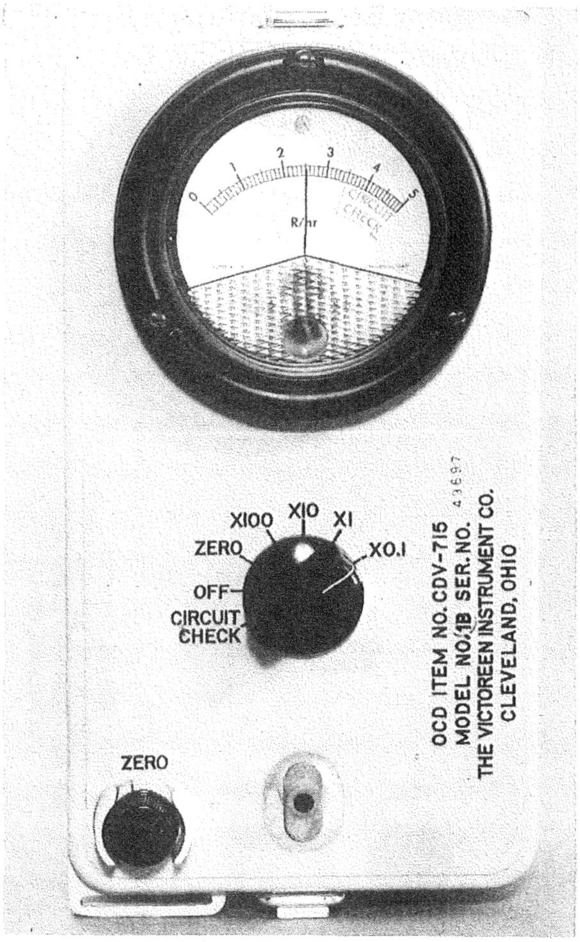

Figure 3-5. The survey meter dial reading is 2.5. This reading is multiplied by 0.1 to get the radiation exposure rate reading because the range-selector switch points to "X0.1." The radiation exposure rate is 0.25 R/hr.

(2.5 x 0.1 = 0.25)

multiplier, the needle will move more for a given change in radiation rate, so you will be able to detect this change easier. For example, the range-selector switch in Figure 3-3, page 3-5, is set at "X10" and the dial reading is only 0.4, for a radiation exposure rate of 4 R/hr. A more accurate reading of 4.1 R/hr is obtained for the same situation by switching the range-selector switch to "X1" as shown in Figure 3-4, page 3-6, where the dial reading is 4.1.

(5) If the needle climbs past 5, the range-selector switch should be switched to a higher range (counterclockwise) until the needle remains on the scale.

e. Troubleshooting the Survey Meter. If you have trouble with the survey meter, it will probably be due to a poor battery, faulty battery installation, or poor battery contacts. Spare batteries should be kept in the shelter. With a good, new battery properly installed, the survey meter should have an operating life of about 200 hours under normal operating conditions. Dirty or corroded contacts can be cleaned with a pencil eraser, steel wool, or by very carefully scraping the contact surfaces with a knife. Bits of eraser, dirt, or steel wool must be very carefully and thoroughly removed from inside the case.

If radioactive dust gets on the outside of the survey meter, it should be carefully cleaned off with a cloth dampened in a mild soap solution. Instruments can be kept in a plastic bag to prevent contamination. If the inside of the meter accidentally becomes contaminated, the instrument should be taken to a clean area where the inside of the bottom case may be carefully cleaned off with a cloth dampened in a mild soap solution. It must be THOROUGHLY DRIED before putting the case together. The electronic components mounted on the inside of the top cover may be brushed and dusted off with a dry brush and/or blown out with dry air. A damp cloth should NOT be used on any of the electronic components. If the remaining interior contamination causes a slightly increased reading only on the "X0.1" range, the instrument will still be useful.

Do not try to make any calibration adjustments or any repairs on the survey meter. Special equipment and specially trained people are necessary to do these jobs.

3-3. Preparation for Using the Dosimeter Charger.

a. Preliminary. The dosimeter charger (Figure 2-3, page 2-5) is necessary to move the hairline on the dosimeter back to the starting (zero) position, as described briefly in paragraph 2-5. Without a dosimeter charger, the dosimeter can't be used after the hairline has reached the end of the scale. Because the dosimeter charger is necessary to use the dosimeter, preparation of the charger is described first.

The charger has one control knob, called the voltage control. When the charger is turned so the printing on the top can be read, this knob is located on the top, far-right corner of the charger. On the top left corner there is a cap with a chain coming out of the top of it; beneath this cap is the charging contact. The chain keeps the cap from getting lost when it is unscrewed and lifted off the charging contact. The cap should be kept screwed down over the charging contact when the charger is not in use, to keep the contact clean, to prevent mechanical damage, and to prevent accidental discharge of the battery.

b. Installing the Battery in the Charger. The charger is powered with a single D-cell flashlight battery. The battery powers two things in the charger: the electronic circuit that charges the dosimeter, and the light bulb that illuminates the dosimeter scale while charging it. CAUTION: THE CHARGER SHOULD NOT BE USED AS A FLASHLIGHT. Using the charger as a flashlight will quickly run down the battery and then, unless you have spare batteries, you won't be able to charge the dosimeters. The battery should be removed if the charger will not be used for a few days or longer.

The battery is installed as follows:

(1) Use a coin or screwdriver to unscrew the large screw at the center of the bottom* of the charger, as shown in Figure 3-6, page 3-9. After a few turns counterclockwise, you will feel that the bottom of the charger case is no longer attached to the top. The screw will not come out and remains attached to its part of the case.

(2) Lift the bottom case (which is now on top) up from the top part.

(3) Install the D-cell battery in the rectangular battery holder mounted on the inside of the top cover. The insides of a charger are shown in Figure 3-7, page 3-9, with the battery inserted in the top part shown at the left. If you have trouble deciding how the battery should be put in, read paragraph 3-2b(4).

(4) Notice the rubber pad glued to the inside floor of the bottom. (The rubber pad may not be in some chargers.) Place the bottom part of the case over the top so the rubber pad is over the battery, and tighten the screw by turning it clockwise.

c. Checking the Battery and the Dosimeter Charger (Operational Check). This procedure is also used for resetting or zeroing a dosimeter. A dosimeter is needed for a full operational check of the charger.

CAUTION: If the dosimeter has been in use to measure radiation dose, you should write down its reading (see paragraph 3-4d) before using it to check the charger. Otherwise, if the charger bulb doesn't light up, you

---

*On some models the screw head is on the top.

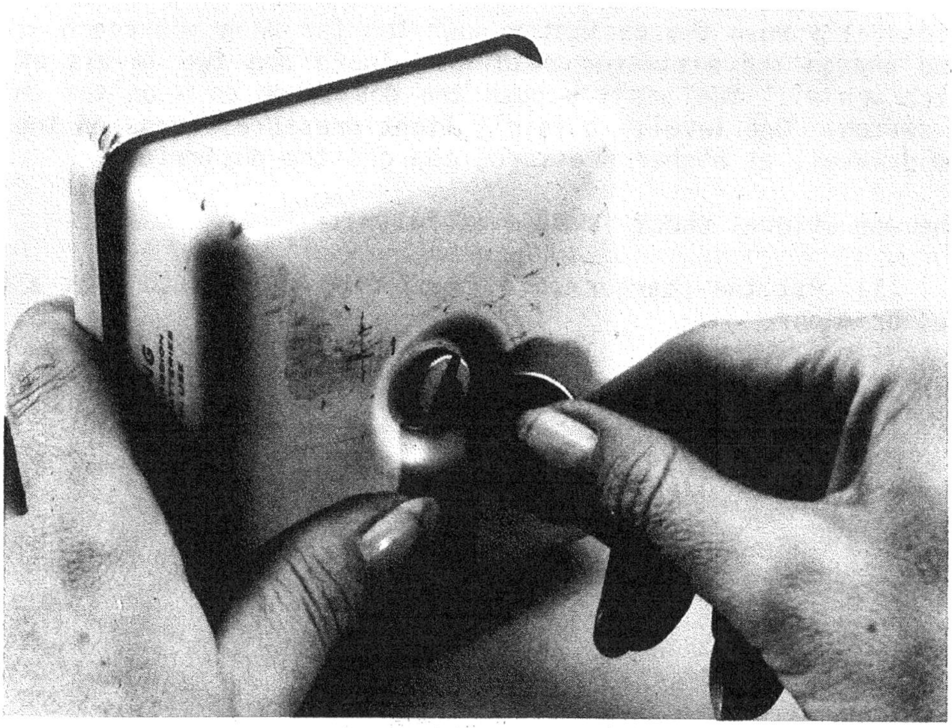

Figure 3-6. A coin may be used to open the dosimeter charger to put in a battery.

Figure 3-7. The interior of a dosimeter charger with battery in place in the inside top cover. The bulb which lights to view the dosimeter is at the upper left, and an extra bulb is mounted just to the left of the battery.

may accidentally push the dosimeter down too far when you reach step (4) below and change the dosimeter reading. There are two levels of pressure which you will feel as you push the dosimeter down on the charging contact switch. One level, at fairly light pressure, turns on the lamp. The second level, at higher pressure, charges the dosimeter.

The operational check is made as follows:

(1) Put the charger on a firm, flat surface such as a table, desk top, or floor.

(2) Unscrew the cap from the charging contact and lay it to one side, as shown in Figure 3-8.

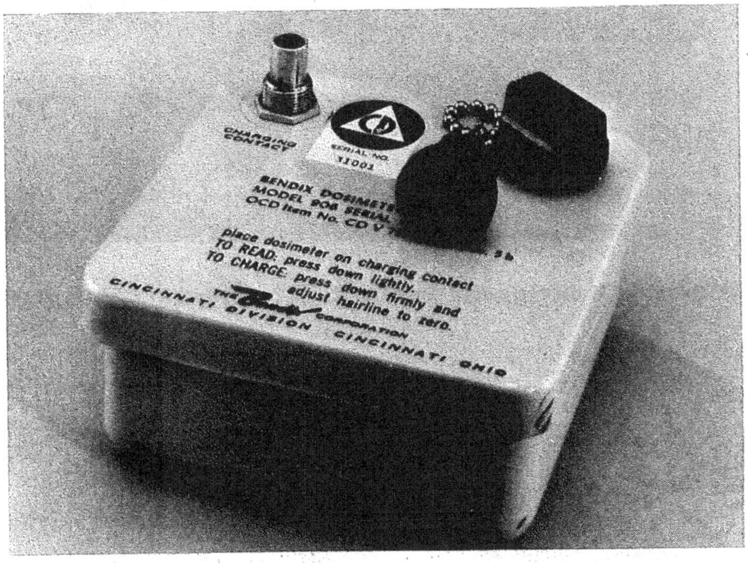

Figure 3-8. A charger with the knob removed from the charging contact.

(3) Place the charging end of the dosimeter over the charging contact, as shown in Figure 3-9, page 3-11. The charging end is opposite the end with the pocket clip and is hollowed out with a center post down inside. You will need to use one hand to hold the dosimeter down on the charging contact and the other hand to adjust the voltage control. You may need to experiment to find out which arrangement of your hands is easiest for you to do the job. If the right hand is used on the dosimeter, you will need to rotate the charger so the printing is away from you, as shown in Figures 3-9 and 3-10, page 3-11.

(4) Look through the dosimeter eyepiece on the end by the pocket clip, and push the dosimeter down gently on the charging contact against the spring pressure until you can see the dosimeter scale light up. If the charger light doesn't come on, check the battery, light bulb, and contacts as described in paragraph 3-3d, "Troubleshooting the Dosimeter Charger."

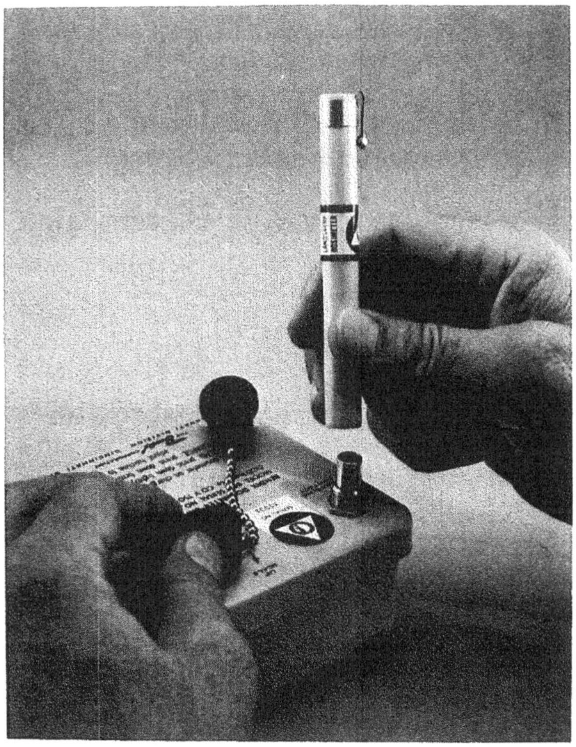

Figure 3-9. Placing a dosimeter on the charger.

Figure 3-10. Resetting a dosimeter to "zero" with a dosimeter charger.

(5) Push down the dosimeter with greater pressure on the charging contact until it reaches bottom and won't go any farther. Hold it there.

(6) While the dosimeter is being held solidly down on the charging contact with one hand, use the other hand to rotate the voltage control knob. Look through the eyepiece to watch the hairline, as shown in Figure 3-10, page 3-11. The hairline should move as you rotate the voltage control, and you should be able to make it move to the "0" (zero) at the left end of the dosimeter scale. If you can't make it move to the zero, check the next section, "Troubleshooting the Dosimeter Charger."

(7) Remove the dosimeter and replace the cap over the charging contact.

d. Troubleshooting the Dosimeter Charger.

(1) Always keep the protective cap on the charging contact when the charger is not in use. The smooth surface of the clear plastic insulator around the center post of the charging contact should be dry, clean, and without fingerprints. Use a soft cloth (free of grit, dirt, lint, and moisture) to clean it. Don't use strong solvents or cleaning fluids to clean plastic parts because some of them can dissolve plastic.

(2) Take out the battery and keep the case closed when the charger is not to be used for periods of several days or longer.

(3) If the light does not come on when the dosimeter is pressed down on the charging contact, do the following:

(a) Check the battery to be sure that it is installed with the correct polarity (in the right direction) and that it is making good electrical contact. If the condition of the battery is questionable, replace it with a battery that is known to be good.

(b) Check the light bulb to see if it is loose in the socket (see Figure 3-6, page 3-9) and tighten if necessary.

(c) Replace the bulb with the spare if there is any chance that the bulb is burned out.

(d) After taking the above actions, if the light still does not come on when the charging contact is depressed, the charger should be returned for repair or replacement, if possible.

(4) If the light is dim or appears weak, do the following:

(a) Check the battery to make sure that good electrical contact is being made.

(b) Clean the battery and light switch contacts with a pencil eraser or steel wool until the metal making contact is bright and shiny.

(c) Tighten the nut on the charging contact.

(d) If the condition of the battery is questionable, replace it with a battery that is known to be good.

(5) If the dosimeter scale is illuminated, but when the voltage control knob is rotated the hairline does not appear on scale, or the hairline is unsteady (jittery movement of the image), do the following:

(a) Check for dirt or moisture on the charging contact or on the charging end of the dosimeter, and clean it off.

(b) Check for good electrical contact between the dosimeter and the outer aluminum sleeve of the charging receptacle. Press the dosimeter down firmly against the charging receptacle and rotate the dosimeter back and forth a half-dozen times. Keeping the dosimeter vertical, move the dosimeter sideways to make the charging contact sleeve touch the inside wall of the dosimeter charging receptacle.

(c) Check for proper electrical contact between the light switch spring contacts, and clean them if necessary (see above).

(d) Try another dosimeter.

(e) If the hairline image still cannot be made to appear on the scale after taking the above actions, the charger should be returned for repair if possible.

3-4. Preparation for Using the Dosimeter

a. Preliminary. What the dosimeter measures and how it works are briefly described in paragraph 2-4. The dosimeter (Figure 2-2, page 2-3) has no battery to install and run down and no controls to operate. As long as the hairline is on the scale when viewed through the eyepiece, the dosimeter can be considered to be turned on. It actually operates continuously. The position of the hairline on the scale can be read anytime and as often as you wish. If the hairline can't be seen, then the dosimeter is useless and must be recharged.

b. Charging or Zeroing the Dosimeter. An electric charge must be placed inside the dosimeter to make the hairline visible and to reset it to the zero position on the scale. A dosimeter charger is necessary for this operation. Exactly the same procedure is used to zero or reset the hairline of the dosimeter as is used for the operational check of the dosimeter charger.

In a fallout situation, be sure to write down the reading on the dosimeter scale, as well as the time, just before the dosimeter is charged or reset to zero. The reason for keeping such records and how to do it are described in paragraph 4-4f, "Keeping Track of Everyone's Radiation Exposure." If you use the charger to read the dosimeter, be careful not to press the dosimeter down too hard on the charging contact or else you will wipe out the reading. Because of this possibility, you may wish to use another light source for reading (when not zeroing) the dosimeter.

The dosimeter is zeroed as follows:

(1) Put the charger on a firm, flat surface such as a table, desk top, or floor.

(2) Unscrew the cap from the charging contact and lay it to one side, as shown in Figure 3-8, page 3-10.

(3) Place the charging end of a dosimeter over the charging contact, as shown in Figure 3-9, page 3-11. The charging end is opposite the end with the pocket clip and is hollowed out with a center post down inside. Use one hand to hold the dosimeter down on the charging contact and the other hand to adjust the voltage control. You may need to experiment to find out which arrangement of your hands is easiest for you to do the job. If the right hand is used on the dosimeter, you will need to rotate the charger so the printing is away from you, as shown in Figures 3-9 and 3-10, page 3-11.

(4) Look through the dosimeter eyepiece on the end by the pocket clip, and push the dosimeter down gently on the charging contact against the spring pressure until you can see the dosimeter scale light up. If the charger light doesn't come on, check the battery, light bulb, and contacts as described in paragraph 3-3d, "Troubleshooting the Dosimeter Charger."

(5) Push the dosimeter with greater pressure down on the charging contact until it reaches bottom and won't go any farther. Hold it there.

(6) While the dosimeter is being held solidly down on the charging contact with one hand, use the other hand to rotate the voltage control knob, and look through the eyepiece to watch the hairline, as shown in Figure 3-10, page 3-11. The hairline should move as you rotate the voltage control, and you should be able to make it move to the "0" (zero) at the left end of the dosimeter scale. If you can't make it move to the zero, check paragraph 3-3d, "Troubleshooting the Dosimeter Charger."

(7) After you have zeroed the hairline, lift the dosimeter from the CHARGING POSITION (which is all the way down) to the VIEWING POSITION (which is almost all the way up), and check the position of the hairline. It may have drifted to one side or the other of the zero, and you will need to zero it again. After a little practice, you will be able to zero the hairline quickly in one try.

(8) Remove the dosimeter and replace the cap over the charging contact of the charger.

c. Checking Dosimeters for Leaks. Dosimeters are very reliable and rugged, but there may occasionally be one which may "leak;" that is, the hairline will slowly drift up-scale to the right of zero, even though there may not be enough radiation around to make the needle move at all in a nonleaker. Most of the leakers should have been weeded out or repaired while they were in storage, but there remains a small chance that you may have a leaking dosimeter in your shelter. If there is time during a crisis period before a nuclear attack, check your dosimeters for leakage as follows:

(1) Zero all dosimeters. Record their serial numbers and the time they are zeroed.

(2) Place the dosimeters in a secure place.

(3) Check each dosimeter and record the readings every 12 hours. You may wish to check them in a shorter time if you think a nuclear attack may happen at any moment. Do not wait until after a nuclear attack has begun to check the dosimeters for leakage. Record the readings and the time even though you check the dosimeters in intervals of less than 12 hours. If a nuclear attack doesn't begin, continue to check the dosimeters for four days (96 hours).

(4) At the end of the leak-checking period, whether four days or less (depending on the situation), calculate the leakage per 24-hr day for each dosimeter. For this calculation, use the final reading on the dosimeter at the end of the leak-checking period. Ignore dosimeter readings taken at other times during the leak-checking period. Multiply this FINAL reading by 24 and then divide by the total number of HOURS in the leak-checking period.*

For example, if you estimate the dosimeter reading to be 2 R (2 roentgens) after a leak-checking period of 8 hours, the leakage rate, L, is $L = (24 \times 2) \div 8 = 6$ R/day.

If a dosimeter leaks as badly as the dosimeter in this example, it can still be used, but you must calculate the leakage and subtract it from the dosimeter reading to get a correct radiation exposure reading. If there is not time or opportunity during a crisis period to exchange dosimeters that leak more than 2-3R per day, they should be marked with

---

*This calculation can be more directly specified by a formula as follows: Let R represent the reading of the dosimeter at the end of the leak-checking period, and let T represent the total number of HOURS in the leak-checking period. The leakage rate per day in roentgens per day, represented by L, is calculated from the formula: $L = 24\ R/T$ (The slash, /, means that the product, 24 R, is divided by T).

an "L" on the body of the dosimeter, either with paint or fingernail polish, if available, or by scratching in the enamel of the dosimeter with a knife or sharp instrument. A label could be attached which shows the leakage rate. The mark or label will alert the person reading the dosimeter not to become unduly alarmed at a high reading on the dosimeter. If the dosimeter leakage is greater than 10 R per day, it should be considered to be unreliable.

d. Reading the Dosimeter. Reading a dosimeter has been discussed in paragraph 2-4, "The Dosimeter," and in paragraph 3-4b, "Charging or Zeroing the Dosimeter." Some additional information will be given in this section.

When you point a dosimeter to a light and look through the eyepiece with your eye about 1/2 inch from the lens, you should see a field of view as illustrated in Figure 3-11, page 3-17. You may need to rotate the dosimeter so the word "ROENTGENS" appears right side up. The hairline is at zero in this illustration, where it should be placed whenever the dosimeter is recharged. In Figure 3-12, page 3-17, the hairline is at about 107 R. If you read the dosimeter with the scale running up and down instead of horizontally, you will get a reading which is slightly wrong, due to the effect of gravity.

The reading of the location of the hairline on the center scale can be estimated to the nearest whole number. For various reasons you may wish to record the dosimeter reading to the nearest whole number, although the accuracy of the instrument is rated at within plus or minus 20 percent when measuring gamma radiation from fallout. This accuracy specification means that if the actual exposure is 107 R, then the dosimeter should read between 86 R and 128 R. There are a couple of reasons for reading and recording the dosimeter to the nearest whole number. In principle, the dosimeter could be calibrated at a later date, if necessary, and the recorded readings might be corrected to a more accurate value. Another reason for recording the reading to the nearest whole number is that you may be interested in seeing small increases in the radiation exposure, from 18 R to 20 R, for example.

You need a light to read the dosimeter. A match, a candle, or a flashlight will do. However, the brighter the light, within reason (the sun is TOO bright), the easier it is to see the scale and the hairline. The dosimeter charger has a built-in light for reading the dosimeter, but the charger must be used with CAUTION. If the dosimeter is pressed down too hard on the charging contact, the dosimeter reading will be lost. Also, use of the charger light will run down the battery unnecessarily if other light sources are available.

When gamma radiation is present, the hairline of the dosimeter is moving all the time, usually so slowly that the motion can't be seen. If the radiation exposure rate is very high, the movement will become visible. For example, if the radiation exposure rate is 7200 R/hr, the hairline of the dosimeter would march across the scale at the rate of

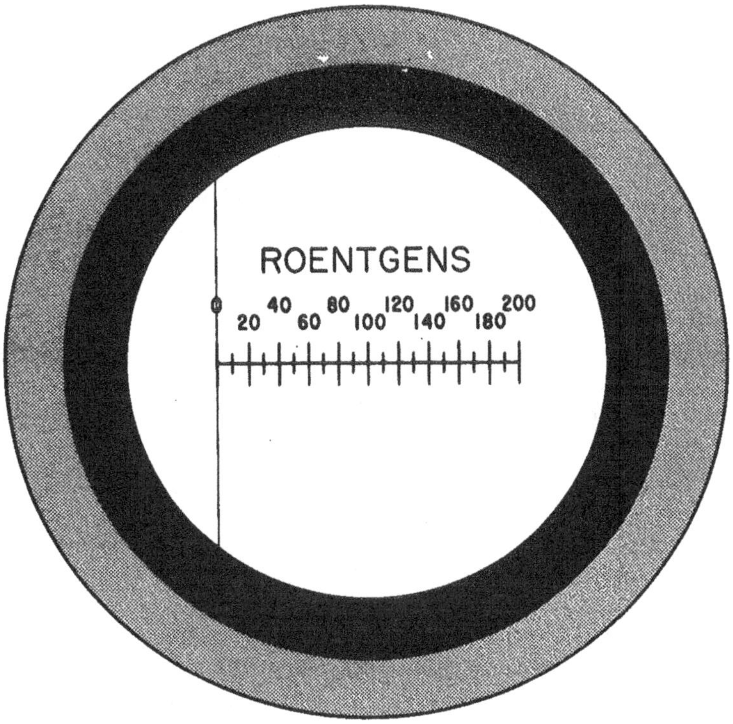

Figure 3-11. Field of view of dosimeter with hairline set at zero.

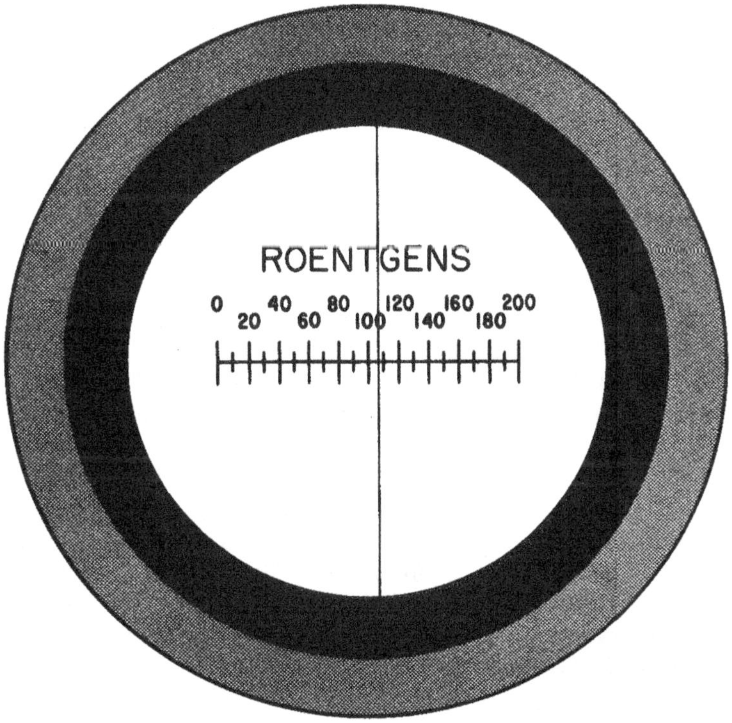

Figure 3-12. Field of view seen in a dosimeter with hairline at about 107 R.

2 R every second. This motion would be quite apparent, but it would be very unhealthy to stay and observe the motion for more than a few seconds!

A dosimeter need not be zeroed to measure the radiation exposure for a specific period of time or a particular mission. Write down the readings taken from the dosimeter before and after the exposure. These readings are called the INITIAL and the FINAL readings. Subtract the initial reading from the final reading (which will always be larger) to get the exposure. For example, if the dosimeter reading is 28 R when the mission is started and 52 R when the mission is completed, the radiation exposure is obtained by subtracting 28 R from 52 R to get 24 R.

This same procedure may be used to estimate radiation exposure RATES if no survey meter is available. Record the reading on the dosimeter (or reset it to zero if desired) and the time, then place the dosimeter in the location where the radiation rate is to be measured. Don't stay with the dosimeter. You won't want the exposure, and you won't want to affect the reading by the shielding that would be caused by your body. To produce a moderately accurate result, the radiation rate should be high enough to cause a change of at least 10 R in a reasonable period of time, say in 5 to 30 minutes if the measurement is being made during the first few days after fallout arrives. You may have to go to the location and read the dosimeter a few times before it shows a suitable change. When the dosimeter reading shows an increase of at least 10 R, record the new reading and the time, and remove the dosimeter from the location. Calculate the radiation exposure rate in roentgens per hour as follows:*

(1) Find the total exposure by subtracting the initial dosimeter reading from the final dosimeter reading.

(2) Multiply this number (the total exposure) by 60.

(3) Divide your result by the total time in MINUTES during which the dosimeter was exposed.

In the first few hours after fallout arrives at a location near the explosion, the radiation rate will decrease rapidly due to the decay of radioactivity. If the dosimeter is used to estimate the radiation exposure rate during this period, the actual radiation exposure rate may be significantly lower by the time the calculation is made.

---

*A formula for this calculation is defined as follows: Let D represent the total exposure in roentgens, obtained from the CHANGE in dosimeter readings, and let T represent the total time of exposure in minutes. Then the radiation exposure rate, R, in roentgens per hour is calculated from the formula: $R = 60\ D/T$.

CHAPTER 4

RADIATION SAFETY PROCEDURES

4-1. Introduction. The first three chapters have given you some facts about nuclear radiation, how it is detected with radiological instruments, and how to operate the civil defense radiological instruments provided for shelters. This chapter tells you how to use that information to provide the greatest possible protection from nuclear radiation while you are in shelter.

4-2. Before Fallout Arrives. In some localities there may be detailed planning and preparation for protection in case of a crisis or emergency during which a nuclear attack might take place. In those localities, many of the tasks described here will already be done before the crisis happens. Even in those localities where as much has been done as possible before a crisis, there will still be some tasks that should be done soon after a crisis occurs.

It may not be possible to do all these tasks before fallout arrives, and in that case, those tasks that can be done inside the shelter can be done later while fallout is arriving. Those tasks that require trips outside the shelter will have to be postponed or forgotten if they are not completed by the time fallout begins to arrive, unless special circumstances of extreme urgency or very low risk make the trips worthwhile. No one who is in a shelter when fallout begins to arrive should leave the shelter except under special circumstances of extreme urgency or very low risk.

If a crisis develops quickly, leading to a nuclear attack on short notice, shelters in the communities where people live would be used. In this case, there probably would not be time to do some tasks before fallout arrives, such as checking the dosimeters for leakage or improving the radiation safety of your shelter. On the other hand, you may know many of the people in the shelter and may have an idea who might be able to help with radiation monitoring and other tasks. You may also know where useful and vital supplies are located. Furthermore, you may be familiar with the shelter and will not need to spend much time checking it out.

If a crisis develops gradually, there may be time for people in high-risk areas, areas which might be targets, to relocate to areas of lower risk. In this case, people who relocate may set up housekeeping in or near the shelter they would use if it became necessary. There would probably be time to work out an organization of the shelter population, check out the shelter, get supplies for maintaining radiation records, stockpile materials for possible use as emergency shielding, and to leak-check the dosimeters.

A discussion of things to do for radiation safety before fallout arrives is given in paragraphs 4-2a through 4-2d. Two checklists for the RM are provided at the beginning of this handbook: Checklist "A" for immediate action (yellow pages), and Checklist "B," a standard checklist for RMs (blue pages).

a. Organization of Shelter Population. The Shelter Manager and assistants will supervise the organization of the shelter population into small groups called shelter units. Organization of the shelter population into shelter units, each with its own Unit Leader, is necessary not only for good management but also for keeping a radiation exposure record for each person in the shelter. There may be between seven and 15 people in a shelter unit. There probably won't be enough dosimeters for each person to have one. The shelter Unit Leaders can help estimate the radiation exposure of those people in their units who don't have dosimeters. The Unit Leaders can also see that someone fills out the radiation exposure record for those who are unable to do it themselves, such as small children.

Organization of the shelter population into shelter units will also be necessary in case people need to be moved to a different location in the shelter where the exposure rates are lower. Unit Leaders can supervise the movement to see that their units move as a group and that no one accidentally moves into a hazardous area.

After the shelter units have been organized and the Unit Leaders selected, the Unit Leaders should be shown how to fill out the radiation exposure records. If blank forms (see Figure 4-1) are available, these should be issued before fallout arrives. The Unit Leader should see that the top part of each form is filled out for everyone in the unit.

Radiation sensitivity categories are listed and described in Table 4-1, page 4-4. Identifying people according to these categories before fallout arrives may be useful if it should later become necessary to arrange for special shielding. The effect of a given whole-body exposure to radiation will vary somewhat among individuals, due partly to age, sex, body thickness, and general health. The sick, aged, and very young are the most susceptible. Nevertheless, it is generally advisable for shelter management to consider the entire shelter population to be equally susceptible to the effects of radiation, with the possible exception that children and pregnant women should be treated as being more susceptible. If a woman is pregnant, her radiation exposure record form should be marked "PG" on the line following "Rad. Sensitivity Category."

RADIATION EXPOSURE RECORD

Name ______________________________

Home Address ______________________________

Social Security No. ______________________________

Shelter Address ______________________________

____

Name of Shelter
Unit Leader ______________________________
Rad. Sensitivity
Category ______________________________

Hour and Date	Added Exposure (R)	Total Exposure To Date	Comments

FRONT SIDE

Hour and Date	Added Exposure (R)	Total Exposure To Date	Comments

BACK SIDE

Figure 4-1. Front and back side of a form for keeping track of the individual's nuclear radiation exposure.

Table 4-1. Radiation sensitivity categories

Category	Description	Cause for Immediate Concern[a]
PG	Pregnant women	Miscarriages, malformed babies, radiation sickness
Child	Infants, small children	More susceptible to radiation injury than adults
Y/A	Youths and adults	Radiation sickness

[a] In addition to radiation sickness, there may be radiation effects that occur many months or years after exposure such as cancer, leukemia, sterility, cataracts, and genetic injury. The probability of developing such late effects should not be a principal determining factor in decision-making during a nuclear war emergency, but such effects can and should be minimized by keeping controllable exposures as low as practicable.

b. Checking Out the Shelter. Many different kinds of shelter will be used for protection against fallout in an emergency. Some shelters may be in schools, churches, or banks. Others may be in factories, office buildings, large stores, underground garages, basements of apartments or houses, mines, or caves. Some shelters may have many rooms, some of them on different levels, and others may have just one large room. The problems of providing the best radiation safety will be a little different in each shelter.

The Emergency Operating Center (EOC) should be consulted if special problems, not discussed in this handbook, should arise. Finding a solution for some of these problems may mean the difference between life and death for some of the people in your shelter. These solutions may depend on how good you are at inventing and putting together ideas on the spot and being able to do things in a difficult situation.

Here is a list of items to check out and do in your shelter before fallout arrives. In the sections following the list, each item is discussed in greater detail. The most urgent items are included in Checklist "A" for immediate action (yellow pages). All of the items are included in Checklist "B," the standard checklist for RMs (blue pages). You, the RM, will have to work in cooperation with the Shelter Manager and others on many of these items.

• Which locations appear to offer best protection against fallout? Sketch a shelter floor plan and mark these locations.

o Is there going to be enough room for all of the people in the location of best protection?

- Can the radiation safety of the shelter be improved with tools, materials, and manpower on hand?

- Are there openings to be baffled or covered to reduce the amount of radiation coming through? Will these changes allow enough air to flow through to keep people from getting too hot when they are crowded?

  o Are materials and tools handy which could be used for putting up additional, improvised shielding inside the shelter after fallout arrives?

  o Is there going to be a problem if a lot of people enter the shelter while fallout is coming down? Are brooms and dustpans on hand to sweep up fallout particles?

- Will trips for water or to restrooms increase radiation exposure?

- Where could dosimeters be mounted or hung? Are needed materials available for mounting or hanging them?

- Where can instruments, instrument supplies, flashlights, and batteries be stored securely?

- Are there enough candles, lanterns, flashlights, and other light sources so you can move around and read instruments if the power goes out?

- Are writing supplies available, including pens or pencils, and printed forms or paper, for keeping records of radiation exposure? Do you have a notebook in which to keep a record (RM log) of events?

(1) Best Protection. Which locations appear to offer the best protection against fallout? Sketch a shelter floor plan and mark these locations.

The best protection is provided by getting as much mass as possible between you and the fallout. Walk through the shelter and get an idea where the best protected areas might be. Usually, but not always, the areas having the least amount of daylight reaching them will provide the best protection.

Basements provide good protection from the sides if they are well below ground and there is earth all around, but they may not always provide good protection from "skyshine" or from radiation from fallout that has settled above the basement or on neighboring rooftops. If the floors above the shelter are made of solid concrete, they will be much more massive than floors of wood and will provide much better protection from overhead gamma radiation. Similarly, walls of solid brick or concrete will provide better protection than walls of hollow concrete or cinder-block; these walls, in turn, will provide better protection than walls of gypsum board or plywood.

Tall buildings can provide good protection from gamma radiation in the inner rooms of floors that are at least four stories above the ground or surrounding rooftops. There should be at least three stories above the shelter to provide protection from fallout on the roof. These locations do not provide blast protection and should not be used in areas less than 25 miles from a likely target for a nuclear weapon.

If we expect the gamma radiation from fallout to be reduced at a certain location by a factor of four from what the radiation level would be outside above a very large, flat, smooth, open area, covered with the same kind and amount of fallout, we say the FALLOUT PROTECTION FACTOR (FPF) of that location is four. This factor is also called the PROTECTION FACTOR (PF). Some locations that are rated with a high protection factor, such as shelters in the upper levels of a skyscraper, may provide little protection against other nuclear weapons effects such as blast. A high FPF for a shelter location only indicates good protection against gamma radiation from fallout. Such a shelter location may also, but not necessarily, provide protection against other nuclear weapons effects. We will use the term FPF in this handbook instead of PF to indicate the protection provided by a shelter location against gamma radiation from fallout.

Some FPFs that might be possible in different locations in buildings are shown in Figure 4-2.

The Shelter Manager may have a sketch of the shelter floor plan or may make arrangements to have one drawn. The sketch should show roughly how the rooms are arranged, the approximate sizes of the rooms, where windows and doors are located, and, if possible, what kinds of materials are used to make the walls. You will use this sketch to keep track of your radiation measurements at different locations, where most of the people are located at different times, and where you might have to construct special shielding. You might ask someone in the shelter to draw or trace several copies for you so you will have a copy for each set of information, circumstances, or instructions.

A sample sketch of a basement of an apartment building is shown in Figure 4-3, page 4-8. We will call this make-believe building Erskine Hall, and we will use it for several examples.

The side view is included because it shows how deep the basement sits in the ground and that the ceiling is a concrete slab which provides good shielding against gamma radiation. It is not always necessary to sketch a side view, but you might want to include one to show a particular feature of the shelter.

The sketch in Figure 4-3, page 4-8, shows the location of four drainpipes from roof gutters. If there were moderate to heavy rainfall after fallout, there could be pileups of fallout by the drainpipes that could increase the gamma radiation along the walls on the inside of the shelter opposite the drainpipes.

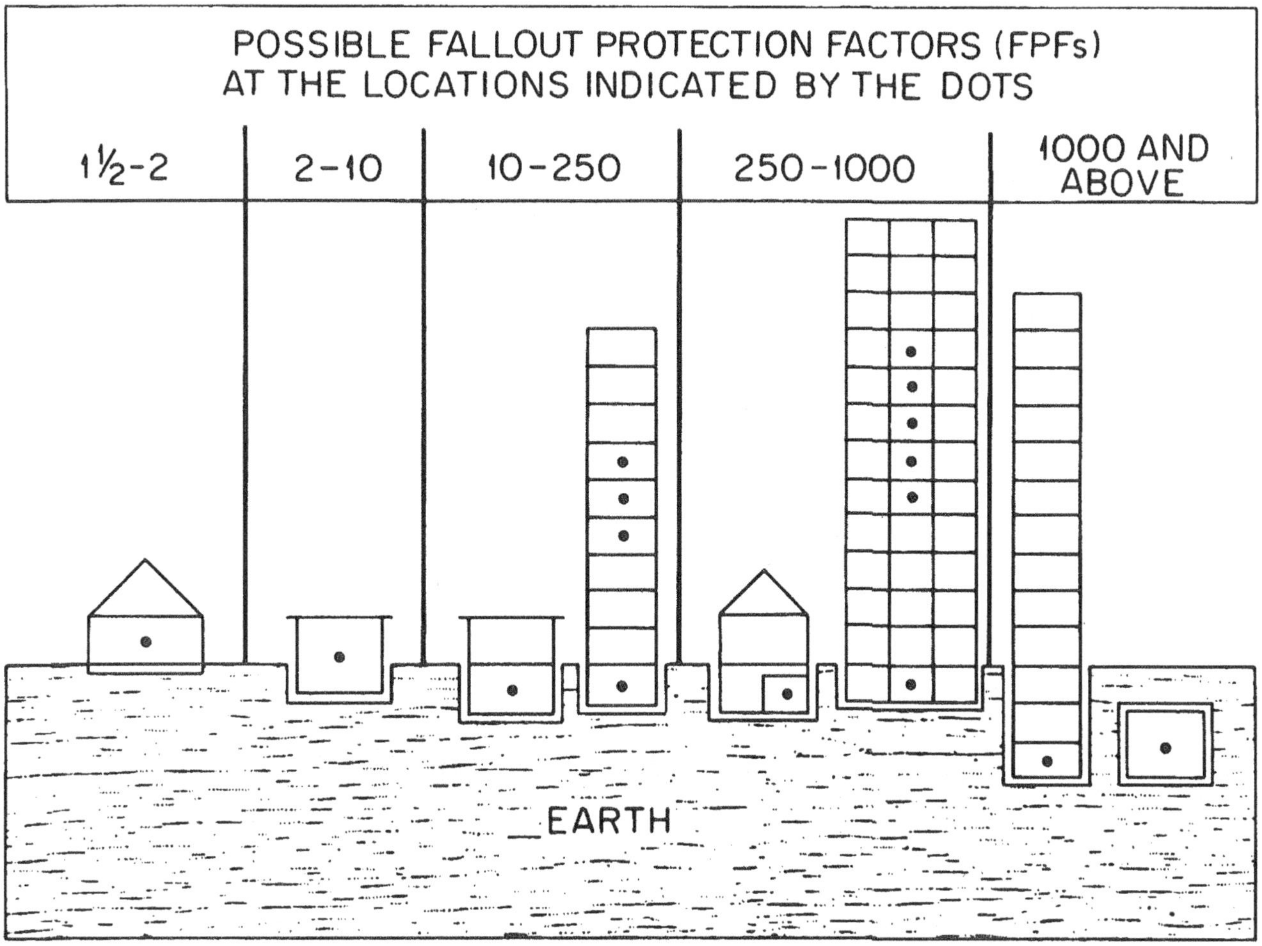

Figure 4-2. Deep basements and buried shelters have high FPFs (1000 and above). They provide good protection against gamma radiation from fallout. Tall buildings also provide good protection against gamma radiation from fallout in the locations indicated by dots in the drawing, but they provide little protection against blast. The FPFs indicated above are for isolated buildings. The FPFs would be higher for ground-level and below-ground shelters that are surrounded or partly surrounded by buildings. The first floors of houses and partially buried basements have low FPFs and provide little protection against gamma radiation from fallout.

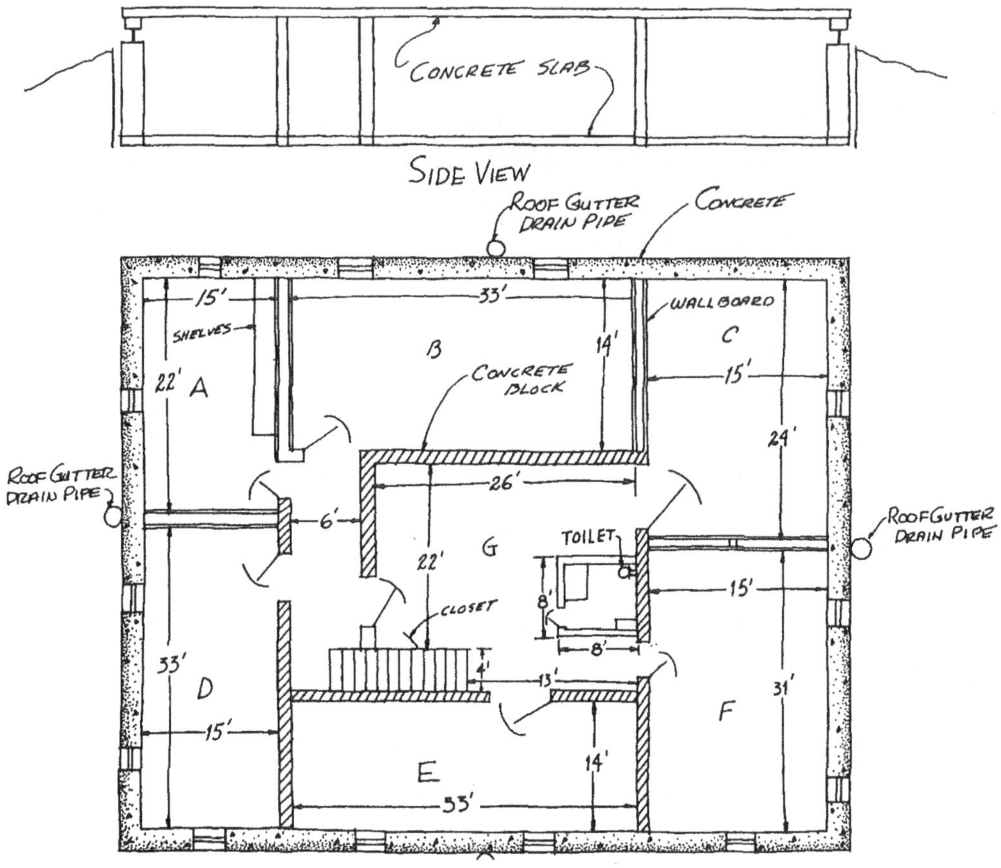

Figure 4-3. Example of a sketch of the floor plan of the basement of a make-believe apartment building called Erskine Hall.

Two kinds of interior wall construction are indicated in the sketch in Figure 4-3, concrete block and wallboard, probably gypsum. The rooms have been named with letters of the alphabet. Room "G" looks like it would provide the highest FPFs because it is surrounded by outside rooms and has walls of concrete block.

(2) Space. Is there going to be enough room for all of the people at this shelter in the locations of best protection?

After the locations have been found that appear to provide the best protection, you should talk with the Shelter Manager about the problem of having enough room. To answer this question you will need to know two things: (a) how many people are in or assigned to your shelter and (b) how much space is available in the locations of best protection.

The Shelter Manager should be able to tell you how many people are already in the shelter or are assigned to it. The Shelter Manager should have a list of names and radiation sensitivity categories (Table 4-1, page 4-4) of occupants, names of shelter Unit Leaders, and a record of kinds of special skills that are available.

To answer the second part of the question, you will need the sketch of the floor plan with the approximate dimensions of rooms. This sketch may not show what is in the rooms. You will need to look at the rooms that you have estimated to be the safest to see if there are furniture, equipment, and obstructions that can be moved to increase the space for people.

Bookcases, boxes, chests, desks, and file cabinets may be moved from the rooms expected to have the highest FPFs into the rooms with lower FPFs. Some kinds of tables should not be moved because people (especially children) may sit under them as well as on top, thus doubling the space. Wide, sturdy storage shelves can also be used for people to sit down or lie upon at more than one level.

If you aren't sure which rooms have the highest FPFs, the Shelter Manager may hold off having items moved until after fallout arrives and the radiation builds up to levels you can detect with the survey meter. Then the survey meter may be used to find the locations with the lowest radiation levels, as described in paragraph 4-4b.

During the early hours after fallout arrives, it may become necessary to crowd people in the safest locations. After the radioactivity decays to a lower level, the occupants can spread out into rooms with relatively higher radiation levels. You can get an idea of whether the Shelter Manager may need to crowd people by estimating the total available space in square feet of the safer locations. Divide that number by 10, the number of square feet allowed per person. If the resulting number is larger than the number of shelter occupants, you have plenty of space in the safer locations. If the number is smaller than the number of shelter occupants, it may be necessary to crowd people temporarily in the safer locations. The number of people in the safer locations can be doubled if you crowd them TEMPORARILY by squeezing down the space per person from 10 sq ft to five sq ft.

In the sketch shown in Figure 4-3, page 4-8, the available floor space in Room G, including the toilet, is about 624 sq ft. The hallway to the left of Room G adds about 132 sq ft for a total of 756 sq ft in the estimated safer locations. Divide 756 by 10, and round off to 76. If more than 76 people are assigned to the apartment building basement in Figure 4-3, page 4-8, they will need to be crowded in Room G and the hallway if the radiation builds up to hazardous levels after fallout arrives. With maximum crowding, they could squeeze about 152 people into Room G and the hallway during the most hazardous times. If more than 152 people were assigned to this shelter, some of them would have to be sheltered in the outer rooms, which are not as safe. In that case, they might work out a rotation scheme so people would share, as fairly as possible, the higher radiation exposures of the outer rooms.

If it is necessary to crowd people in the safer locations, it is very important that enough fresh air and light are provided so that people don't pass out from heat prostration or get claustrophobia (fear of confined, crowded places) and run outside. Both the Shelter Manager and the RM will be involved in these problems.

(3) Radiation Safety Improvement. Can the radiation safety of the shelter be improved with tools, materials, and manpower on hand?

As you go through your shelter looking for the places that appear to provide the best shielding from gamma radiation, you should also look for ways to improve the shielding. Look for openings that can be covered up and for places where walls and ceilings can be thickened to cut down gamma penetration.

In the example shown in Figure 4-3, page 4-8, the radiation safety could be improved with a little effort. Earth could be piled up around the outside where the basement wall rises above ground level. All but two or three basement windows could be sealed with boards or with cardboard and plastic and then covered with earth. The remaining windows may be needed for ventilation and should be baffled rather than sealed. A way to construct a baffle over a basement window to reduce gamma penetration and prevent fallout from entering is shown in Figure 4-4.

About 40-50 manhours of labor would be needed for the improvements in the radiation safety of this shelter. Shovels, picks, and some carpenter's tools (hammers and saws) and supplies (nails, lumber, plywood, plastic sheeting and gloves) would be needed. People who are not accustomed to manual labor should wear gloves from the start when picking or shoveling earth. Blisters are painful and can develop into serious infections, especially if antibiotics aren't available.

These efforts could improve the FPFs of this shelter by factors of four to 10. If the FPF of the safest location were about 25 before these improvements, the FPF could be 100 to 250 afterwards. If the fallout is heavy, this improvement could mean the difference between life and death for the occupants.

(4) Openings and Ventilation. Are there openings to be baffled or covered to reduce the amount of radiation coming through them? Will these changes allow enough air to flow through to keep people from getting too hot when they are crowded?

Both the Shelter Manager and the RM will be involved with the problem of providing enough ventilation while maintaining the best radiation safety, as mentioned in paragraph 4-2b(2).

In the basement shelter of Erskine Hall (sketched in Figure 4-3, page 4-8), all the windows except two or three should be sealed and covered with earth, as discussed in paragraph 4-2b(3). Two or three windows should be left uncovered to provide ventilation. These uncovered windows should be located on the side where fallout is least

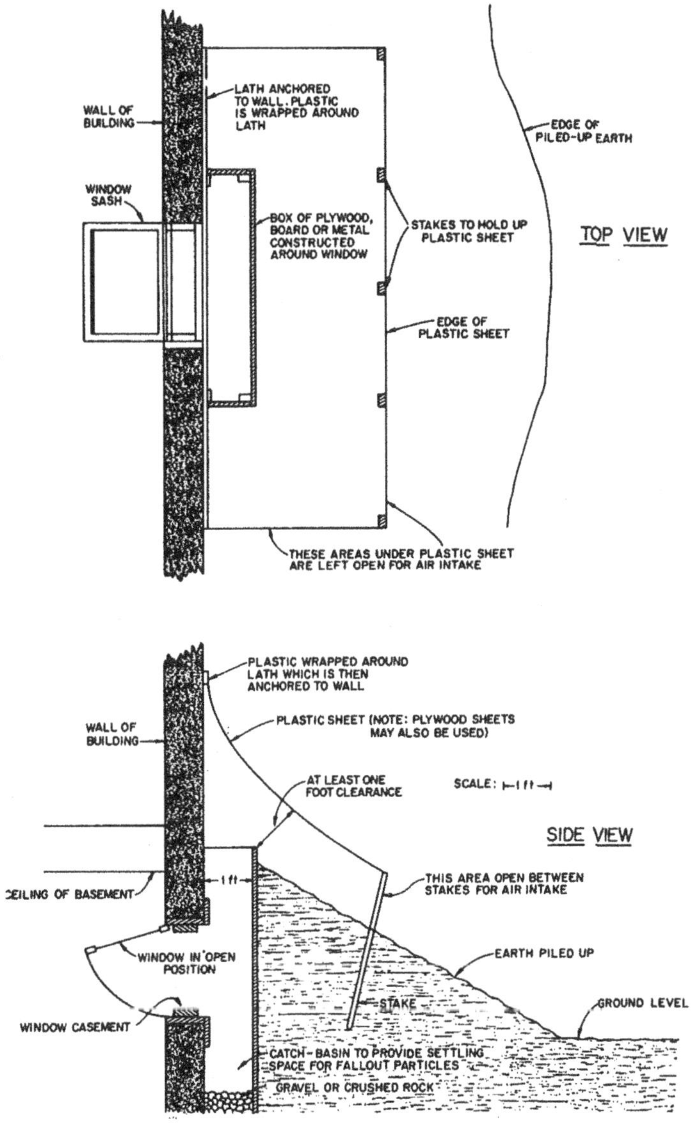

Figure 4-4. A way to construct a baffle over a basement window to allow air to flow in while reducing gamma penetration and preventing fallout from entering.

likely to pile up. If the wind usually blows from the northwest, these uncovered windows should be located on the south or east side. In Figure 4-3, page 4-8, if the top of the figure is north, the uncovered windows should be the two windows near the corner in Room F. If the local wind is blowing from the northwest when fallout is coming down, there may be less radiation buildup at the open windows on the southeast side.

These two uncovered windows should have a baffle or wall built around them with earth piled up on the outside (as shown in Figure 4-4, page 4-11) to reduce the gamma radiation which shines directly into the shelter from fallout on the ground. If the bottom of the window is at ground level, the inside of the baffle should be dug down several inches below the level of the window to provide a trap for fallout particles. If plastic or plywood sheeting is not available, a trough or a pipe from the inside of the enclosed trap to the outside ground level at a lower point is needed to provide drainage.

(5) Materials for Shielding. Are materials and tools handy which could be used for putting up improvised shielding after fallout arrives?

You may have improved the radiation safety of the shelter to the best of your judgment and capability, as discussed in paragraph 4-2b(3). But after fallout arrives, you may find with the use of your survey meter that gamma radiation is shining through at some unexpected location. You should know where and what materials are available to stack up against or cover a wall, doorway, window, or portion of a ceiling to reduce the gamma penetration. Such materials as books, bricks, earth, or wood may be used. Other materials and their shielding effectiveness are listed in Table 1-2, page 1-11. If some of these materials are located outside the shelter, set up (or ask the Shelter Manager to set up) a work crew to move as much of it inside as possible before fallout arrives.

(6) Entranceway Problems. Is there going to be a problem if a lot of people enter the shelter while fallout is coming down?

One problem that could develop is that the shelter entrance could be blocked by people who have stopped just inside the entrance. They may have stopped to brush off fallout particles or, if the shelter is a large building, they may not know where to go.

If there is a possibility of problems at the entrances, one or two people should be selected to be receptionists at each entrance. The receptionists should see that people brush off fallout and shake outer garments if they come to the shelter after fallout begins to come down. Decontamination of people caught in fallout is described in paragraph 4-4a. The receptionists should also show people where to put outer garments from which fallout particles can't be shaken easily, show them where to go in the shelter, sweep or vacuum fallout particles whenever they accumulate, and throw the swept-up particles outside.

The receptionists will need to wear dosimeters and must know how to read them. They should leave the entrance area and go back to the safest part of the shelter as soon as their dosimeters read some preselected limit, such as 10 R. They may leave sooner if no one has arrived after fallout begins to come down.

The receptionists should set up places to store umbrellas, coats, and other outer garments if there are no convenient places to put these articles near the entrances. They should also have brooms and dustpans available.

It may be helpful to tape up sheets of paper near the entrances which show the way to the safest places in the shelter. If there are no receptionists at the entrances, tape up a sheet of paper near the entrances with information on how to decontaminate oneself.

(7) Restroom and Water Locations. Are trips for water or to restrooms going to increase radiation exposure?

The RM should note where drinking fountains, water outlets, and restrooms are located throughout the shelter. After fallout has arrived, he or she should check the radiation levels at these locations. Some of them may have to be blocked off until the radiation decays to a safer level.

In nearly all public fallout shelters, there will be plenty of water for drinking, cooking, and flushing toilets as long as there are no nuclear detonations close enough to break water lines, damage storage tanks, or cause an electric power failure. If the electric power is knocked out by a distant nuclear explosion, there will still be water in the pipes and tanks, which will flow by gravity. Water should be used as needed for drinking and sparingly for other purposes throughout the emergency.

In a nuclear war there is a possibility that the water supply might fail, so water should be stored in the shelter before fallout arrives. If the shelter runs out of water in a heavy fallout area, the RM may be faced with some difficult decisions and unpleasant situations. About two weeks' supply of water should be stored in areas where heavy fallout is expected. About two weeks after fallout has arrived, the radiation intensity even in the worst places will decay to levels where people can make emergency trips without the risk of radiation sickness or death. In areas where heavy fallout is expected and in the case of hot, crowded conditions in the fallout shelters, a minimum of about seven gallons of water should be stored per person, just for drinking.

(8) Dosimeter Locations. Where could dosimeters be mounted or hung? Are needed materials available for mounting or hanging them?

In some shelters where the FPF is high and about the same everywhere, as in deep underground shelters, caves, and mines, only a few dosimeters need to be mounted or hung where people will be located, to get an idea of what total exposures they are getting, if any. Tape, thumbtacks, nails, and string can be used to mount dosimeters.

In shelters where the FPF may change as you move from one location to another, you will need to issue one or two dosimeters to each shelter Unit Leader. The Unit Leader will then be responsible for estimating radiation exposure readings for the members of his or her unit. At certain times of the day or night, the Unit Leader may want to mount or hang one dosimeter in the vicinity of his or her unit and will then need materials for mounting or hanging it.

(9) Instrument Storage. Where can instruments, instrument supplies, flashlights, and batteries be stored securely?

A central and secure location should be found for storing these items. In the shelter sketched in Figure 4-3, page 4-8, the closet under the stairs in Room G can be used. If you can't lock the door when you must leave, find someone to watch over the supplies. Don't let children play with the radiological instruments.

(10) Light Sources. Are there enough candles, lanterns, flashlights, and other light sources to provide light so you can move around and read the instruments if the power goes out?

As mentioned before, electricity may fail in many locations due to a wide-scale nuclear attack. Most of the shelters with the highest FPFs will also have the least daylight reaching them. If the power goes out, these shelters may be pitch black. Some light must be provided so people won't get hurt when they try to move around. You will need a light of some kind to read the radiological instruments. You should have your own flashlight or lantern so you can move around freely and read your instruments whenever necessary.

(11) Writing Supplies. Are writing supplies available, including pens or pencils and printed forms or paper, for keeping records of radiation exposure?

The radiation exposure of each shelter occupant should be recorded every day and for any special trip that increases the person's exposure. A sample radiation exposure record is shown in Figure 4-1, page 4-3, and at the back of this handbook. If enough printed forms for this record are not available, ordinary notebook paper or stationery may be used. If no paper is on hand in the shelter and none is obtainable before fallout arrives, the records may be written on the walls or on whatever materials and surfaces are available.

Remember, the main purpose of the record is to help each person limit their radiation exposure and prevent radiation sickness. If people don't know what they've been exposed to, they won't know whether they are going to get radiation sickness if they make a trip out of the shelter. Each person needs to know their own exposure so he or she can decide whether a trip outside can be safely made.

It will be useful to have a lot of paper to write and draw on in the shelter, not only for radiation records but for shelter sketches, messages, and bulletins. You will need a notebook, which we will call the RM Log, to keep a record of events. In this log you should enter such information as the time and date and a brief description whenever explosions are heard or detected, when fallout arrives, when the peak radiation exposure rates are measured, when and where special measurements are made, and when there is trouble with instruments.

c. Getting and Checking the Instruments. Each county may have a slightly different procedure for getting radiological instruments to the shelters, if they are not there already. In some counties the instruments may be delivered, but in most counties the RM will be expected to pick up the instruments for the shelter. If you are selected to be an RM after you arrive at the shelter, you may have to find out where the instruments are, and you may have to make a special trip to get them. Instructions on how to use the instruments may be given at the place where they are issued. If the RM has not used the instruments recently and no instructions are given, the RM should read Chapters 2 and 3 of this handbook before trying to operate them.

If available, there should be at least one dosimeter for each shelter unit (paragraph 4-2a, "Organization of Shelter Population") and one dosimeter each for the Shelter Manager and the RM. It would be desirable to have one survey meter for approximately every 200 occupants in a shelter and as many dosimeter chargers as there are survey meters. You should get one extra D-cell battery for each survey meter and each charger. If extra batteries are not supplied with the instruments and if there is time, go to a store and buy them.

An operational check on the instruments should be made as soon as they are received, preferably at the place they are issued. Instructions for operational checks are given in Chapter 3 of this handbook.

When you have the instruments at the shelter, go through another operational check. Zero the dosimeters, if they haven't been zeroed already (paragraph 3-4b, "Charging or Zeroing the Dosimeter"). If there is time, start a leak check on all dosimeters (paragraph 3-4c, "Checking Dosimeters for Leaks").

Let the Shelter Manager know that you have the instruments and their condition.

Keep the instruments in a secure place until they are put to use. If you can't lock them up, find someone reliable to watch over them.

d. Informing the People in the Shelter about Radiation Exposure. Many people have a great fear of "invisible death" from nuclear radiation. There will be much anxiety among people in a shelter when it is known that they are getting radiation from fallout. Even if people are frightened, it is better not to hold back information. The policy of "what they don't know won't hurt them" has never worked with the American public.

When the presence of fallout radiation first causes the needle to move up on the survey meter, the people in the shelter should be informed. If there are several people watching the survey meter, the news of fallout radiation will travel very quickly through the shelter.

In order to let people know the radiation levels, select at least one place in each small or medium-sized room where people are sheltered (more places in large rooms) to mount a sheet of paper on which the survey-meter readings taken near the paper will be written periodically. A sample sheet is shown in Figure 4-5, page 4-17. This sheet and the measurements will be discussed again in paragraphs 4-4f and 4-5a.

If there is time before fallout arrives, each shelter Unit Leader should be shown how to read a dosimeter. Each Unit Leader should be encouraged to read the first chapter of this handbook, if they haven't read it already. If there is only one copy, the fastest readers should be the first ones given the handbook to read.

4-3. Watching for Fallout to Arrive. People may find that a nuclear attack is about to happen or is on its way by announcements on the radio or television, by sirens or other warning devices, or by word of mouth. When a nuclear weapon explodes anywhere within several hundred miles, there will be many signs to indicate it. By that time, people should be on the way to, or already at, their shelter. No one should be outside or very far from a shelter when fallout begins to come down.

A nuclear explosion several hundred miles away can cause an electromagnetic pulse (EMP) which may burn out the transmitting capability of some radio and television stations and knock out some telephone circuits. The EMP may also affect power lines, causing momentary blackout or flickering of lights. It may cause a lot of static similar to lightning static in AM radios, and may burn out FM radios or televisions with large antennas. Nuclear explosions near power lines or power stations may cause widespread power blackouts. Nuclear explosions produce a brilliant flash and glow in the sky which may be seen 50-100 miles away in the daytime if the weather is clear, and much farther at night. STARING AT THE FLASH MAY CAUSE EYE DAMAGE EVEN IF THE BURST IS FAR AWAY. A shaking of the ground as in a mild earthquake may follow within a few minutes, depending on the distance from the burst.

The following procedure applies to shelters that are located at least 25 miles away from a likely target for a nuclear weapon. After nuclear explosions have taken place with noticeable effects in or near the shelter, or when notified by the EOC, the RM (for whom the following is written) should take the survey meter outside or by an outside window (on the windward side, if possible) and watch for the arrival of fallout. If the FPF of the shelter is high and the fallout is light in the area, the survey meter may not show that fallout has arrived if the meter is kept at the safest place in the shelter. It is necessary to know when fallout has arrived, even if it is light, so that exposure control measures can be started.

SURVEY-METER READINGS

LOCATION: 1. CENTER, NORTH WALL, ROOM G, ERSKINE HALL

Date	Time	Reading (R/hr)	Comments	Date	Time	Reading (R/hr)	Comments
5 JULY	1020	0.10	FIRST FALLOUT READING	5 JULY	1630	2.75	
"	1030	0.18		"	1645	3.0	CLIMBING AGAIN
"	1045	0.50			1700	3.5	
"	1100	0.87			1715	4.0	
"	1115	1.2			1730	4.5	
"	1130	1.55			1745	5.0	
"	1145	1.95			1800	5.5	
"	1200	2.3			1815	5.75	
"	1215	2.75			1830	6.0	
"	1230	3.0			1845	6.0	
"	1245	3.25			1900	6.2	
"	1300	3.25	LEVELING OFF		1915	6.2	LEVELING OFF
"	1315	3.25			1930	6.2	
"	1330	3.25			1945	6.0	FALLING AGAIN
"	1345	3.25			2000	6.0	
"	1400	3.25			2015	5.75	
"	1415	3.25			2030	5.75	
"	1430	3.25			2045	5.5	
"	1445	3.15	FALLING		2100	5.5	
"	1500	3.10	EXPLOSION HEARD				
"	1515	3.05					
"	1530	3.0					
"	1545	2.95					
"	1600	2.9					
"	1615	2.85					

Figure 4-5. Sample survey-meter readings at location 1 in make-believe Erskine Hall.

If you, the RM, must go outside, keep fallout particles from getting in your clothes and on your skin and hair. Carry an umbrella and wear a hat and an outer garment if available. You should enclose your survey meter in a clear plastic bag, if available, to keep it from getting contaminated. Carry a dosimeter in a breast pocket or on a chain or string around your neck. Take along a transistor radio or a two-way radio, if available, to keep informed of the situation around you. If it is nighttime, take a flashlight along even though the power may be on and the area may be brightly illuminated at the time you start your watch. If fallout is expected to arrive within the hour, zero your survey meter and leave it on with the range-selector switch turned to "X0.1." If fallout is not expected to arrive for an hour or more, leave the survey meter turned off to save the batteries. You may want to turn it on every 10 or 15 minutes just to check the situation.

If fallout arrives from a ground explosion 25-75 miles upwind, depending upon the yield of the weapon, you will probably notice its arrival by the sound of gritty particles striking the window or surfaces around you. You may hear these gritty particles striking for many seconds before the needle on your survey meter begins to climb. When the needle reaches 0.1 R/hr, note the time; enter the shelter; decontaminate yourself (paragraph 4.4a) if you have been outside; record the reading, time, and date in your RM Log; and tell the Shelter Manager and occupants that fallout has arrived. If fallout arrival is to be reported to your EOC, it should be done in accordance with your local plan.

Some people may be working outside the shelter to improve its radiation safety, or they may be carrying shielding materials into the shelter up to the last minute before fallout arrives. They may become aware of the arrival of fallout by noticing gritty particles striking their skin, by hearing them strike nearby surfaces, or by seeing the buildup of particles on surfaces. These people should then go inside the shelter and brush the fallout particles off their clothes and bodies. If they do not notice the arrival of fallout, you, the RM, should tell them that the arrival of fallout has been detected by the survey meter.

If fallout comes to the shelter from many large ground bursts 100 miles or more upwind, the fallout may not arrive for many hours. The fallout may be hazardous even though it arrives as late as 24 hours after the explosions. You may decide not to set up your own watch for fallout for that length of time if your shelter has good two-way communication with the local EOC. If the people in your shelter feel they can rely on the local EOC, they may decide to depend on the announcements from the EOC to let you know how fast fallout is coming to your shelter. These announcements should come at least every half-hour or hour from the EOC, depending on the situation. When it appears that fallout might arrive at your shelter in two or three hours, take the survey meter to a window or outside and begin to watch for fallout.

The people in the shelter may want to have their own lookout for fallout, even though the EOC may seem to be reliable. If you expect the fallout to take a long time to arrive, arrange for people to take turns or shifts in watching for its arrival.

When fallout arrives from distant explosions, you may not notice it as much as you would notice the fallout from closer explosions. The particles may be so small that you may not feel them as they land on your skin. The climbing of the needle on the survey meter may be the only indication that fallout from distant explosions has arrived.

The fallout is carried most of the way to its destination by winds at high altitudes. On some days the wind at high altitudes may be blowing in a different direction from the wind on the ground. Under these conditions, you might think fallout from a particular nuclear explosion will not come your way because the wind where you are is not coming from the direction of the explosion. In this situation, the fallout might arrive at your shelter contrary to your expectations. The direction that the particles are blown by the surface winds may make it seem that they are coming from the wrong direction. Unless you have positive information on the direction the fallout is being carried, do not make any assumptions about where it will come down.

4-4. While Fallout is Coming Down

a. Decontamination of People Caught in Fallout. Fallout arriving within a few hours after a nuclear explosion is highly radioactive. If it collects on the skin in large enough quantities it can cause beta burns (see paragraph 1-8b, "Symptoms of Radiation Injury").

People who are caught outside in fallout should brush fallout particles off themselves and shake out their outer garments as soon as they get inside the shelter. Some people may be carrying umbrellas and wearing raincoats to keep the fallout particles off their skin and hair. If people have not taken such precautions, they should try to get the fallout particles off their skin and out of their hair and clothing as much as possible before going further into the shelter, but they should not block the entrance so others can't get in. It is more important that people get into the shelter than it is to get every speck of fallout off every person before they go further into the shelter. Fallout particles that are carried into the shelter can be swept up and thrown outside.

If there is a possibility of blockage at the entrances because of a lot of people coming to the shelter after fallout arrives, one or two receptionists should be assigned to each entrance to supervise the decontamination. Each receptionist should wear a dosimeter. Arrangements should be made for them to be replaced so they can leave the entrance area as soon as their dosimeters show that they have been exposed to some preselected limit, such as 10 R, of radiation. If only one or two people come every few minutes to the shelter, the receptionists should go back to the safer parts of the shelter. Instructions for decontamination and directions to the safest shelter locations should be printed on sheets of paper and taped or tacked up in places where incoming people can easily see them.

Most fallout particles will be like grains of fine, dark sand and can be easily brushed off from dry surfaces. The particles can be removed from tightly woven fabrics and rainwear by lightly shaking them.

Loosely-woven outer garments such as knitted sweaters, shawls, and scarves may hold fallout particles even after a hard shaking. These garments should be stored in a special place set aside for them until they can be washed. After they are washed, they will be suitable for normal use. The fallout particles will come out in the wash, and the fallout particles or the radiation will not damage the fabric or make it radioactive.

Fallout particles may stick to moist or oily surfaces, including sweaty or oily skin or hair. These surfaces should be carefully wiped or washed off. If contaminated hair cannot be washed, it should be thoroughly brushed or combed, with frequent shaking and wiping of the hair and also of the brush or comb.

It is not necessary to get the last speck of fallout out of the clothing or hair or off the skin. A few grains of fallout carried by each person into the safest parts of the shelter will produce no noticeable increase in the radiation hazard and will not be detectable by the radiological instruments. Daily sweeping of the shelter for hygienic reasons will eliminate most fallout particles that may be carried into the shelter even after decontamination procedures.

The reception area should be organized so people can shake out their outer garments without getting the particles on people around them. After they have shaken out their clothing and wiped off their exposed skin, they should move further into the shelter and sweep the dust off their shoes with a brush or broom. If the shoes are caked with mud or dust, they should be left in the reception area.

Because the fallout particles will fall down to the floor, decontamination of a person should begin with the head and end with the feet. Brushing off or removing the shoes will be the last step of decontamination before a person enters the safer parts of a shelter.

b. <u>Finding the Places with the Lowest Radiation Levels in the Shelter</u>. After the announcement is made to the people in the shelter that fallout has begun to come down outside, you (the RM) should use the survey meter to find the places that have the lowest radiation levels. The people in the shelter should be gathered at the locations that are estimated to have the lowest radiation levels. It should be explained to the people, or at least to the shelter Unit Leaders, that these locations were chosen on the basis of estimates and that places with lower radiation levels might be found by taking readings with the survey meter.

Mark the sketch of the shelter to show the locations where you plan to take readings of the radiation levels. Some of these readings should be taken near walls, posts, or columns upon which you can tape a form showing your readings. An example is shown in Figure 4-5, page 4-17. A general survey of radiation levels with the survey meter should be made as soon as possible after the gamma radiation reaches levels that can be detected inside the shelter. Write down the readings, the times the readings were made, and the exact location for each reading so you can find the same spot when you check later. You may wish to mark the floor where you make your measurements and assign a number to each location.

At this time, when you are trying to find the safest places in the shelter as quickly as possible, you should take readings only in those locations where you estimate the lowest radiation levels will be. For example, if you are in a basement shelter you should not take readings on the first floor at this time. If you are in a skyscraper shelter, there is no need to take readings near an outside wall at this time. The first survey should be spread out to get a general picture of the best shelter areas. Follow-up surveys should then be made to get a detailed picture of radiation levels in the areas where people are finally sheltered.

While fallout is coming down, the radiation levels may be climbing fast. Inside the shelter at the location that you have estimated to be the safest, your survey meter needle may be climbing as fast as one to five smallest divisions on the "X0.1" scale each minute. If you plan to make a detailed comparison between the readings at several locations, the reading at the final location may be quite a lot higher by the time you get to it than it was when you began to take readings. You will not be able to tell whether the higher reading results from a lower FPF or from an increase in radiation levels at all locations of the shelter. The readings would have to be taken in both places at the same time to show which location had the lowest radiation level. You can only be at one place at one time!* You should not wait until the radiation levels stop climbing to make your detailed follow-up measurements, because it might be several hours before the fallout stops coming down. To get a proper comparison of the radiation safety between different locations while the radiation levels are climbing rapidly (due to the buildup of fallout), you will need to use a special method for taking measurements. One of the simplest methods for taking such measurements is the TIME-AVERAGING method described in paragraph 4-4c.

*If your shelter has two or more survey meters (most will not) and two or more RMs, you may work out a simpler method by making readings synchronized by timepieces showing seconds or by the use of two-way radio communication between the RMs. The meters should be compared at one location (identical radiation levels) before AND AFTER the measurement (the instruments may drift) to make sure they read the same or to compensate for different readings.

Another method, to be used if no survey meter is available, is to place a dosimeter at each location to be checked. All the dosimeters to be used should be carefully zeroed at approximately the same time before positioning them. You may have to wait several hours before significant differences in the readings are observed, because the smallest division on the dosimeter is 10 R. With a survey meter, you will be able to compare the radiation levels at several locations within just a few minutes by using the time-averaging method.

c. The Time-Averaging Method. The time-averaging method is used to compare the radiation levels between two or more locations in a shelter when the radiation levels are climbing rapidly and when you have only one survey meter. If only two locations are to be compared and only a few seconds are needed to get from one location to another, the time-averaging method need not be used. The readings obtained at the two locations may be compared directly in that case.

The time-averaging method is a way to estimate what the approximate radiation levels WERE at several locations at ONE particular time. It consists of taking readings at different locations BEFORE AND AFTER one particular time, then averaging those readings to get the reading at that particular time.

If only two locations are to be compared (locations 1 and 2), a reading is first taken at location 1. A short time later, a reading is taken at location 2. After another short period of time OF EQUAL DURATION, whether 30 seconds or one or two minutes, a reading is taken at 1 again. The two readings taken at 1 are then averaged (add them and divide by two) and compared with the reading at 2.

If three locations are to be compared (locations 1, 2 and 3) with equal time intervals of say, one minute between readings, the readings are taken at locations 1, 2 and 3 and then at locations 2 and 1 again, IN THAT ORDER. The order of measurements, 1-2-3-2-1, must not be changed. The two measurements at 2 are taken ONE MINUTE BEFORE and ONE MINUTE AFTER the measurement at 3, the middle or CENTRAL measurement. The two measurements at 1 are taken TWO MINUTES BEFORE and TWO MINUTES AFTER the central measurement. The two readings at 1 are averaged, and the two readings at 2 are averaged to give approximations of what the readings would have been at those locations at the same time that the reading at location 3 was taken.

To use the time-averaging method, you will need a wristwatch or clock that shows seconds as well as minutes. You should have an assistant to help you move quickly through crowds of people, watch the time, and help keep track of the measurements.

Remember that the survey meter does not respond instantly to the radiation it is measuring when the range-selector switch is turned to "X0.1." You will need to allow a few seconds at each location for the needle on the meter to reach its final reading. Do not move, jiggle, or rotate the survey meter while the needle is settling down.

The survey meter should be held about three feet above the floor or at about waist level and about two feet away from the body when taking measurements. If you are taking measurements in a ground-level or belowground shelter full of people, it is important that all the people sit or lie on the floor while you take the measurement. If people are standing, they will shield some of the gamma radiation from your instrument, and your survey meter will then show a lower reading than it would if people were sitting or lying down or if the room were empty. If you used this reading to compare with readings in other locations that are empty, you might conclude that the room with the people in it is safer, although it may actually be more hazardous.

If you plan to compare the readings at several locations, start the first reading where you think the reading should be the lowest, which should be where the people are located. Begin the readings 20-30 minutes after the needle reads about 0.1 R/hr in the safest location, after you have made your first rapid, spread-out survey. If you start in another location, you may find that when you get to the estimated safest location, the radiation level may still not be high enough to read on the meter. You will then have to repeat the measurements later.

The 20- to 30-minute waiting period will allow time for enough fallout to settle on the ground so the readings will not be influenced much by radiation from fallout particles still in the air. You may wish to use this period to choose the exact locations where you will take measurements, mark these locations on your sketch and at the actual spot, and prepare a sheet of paper or a page in the RM log so your measurements can be written in the correct place when you take them. You should have an assistant with you while you make these preparations so he or she will know what to do when you are taking the measurements. An example of the time-averaging method for comparing seven locations is shown in Table 4-2, page 4-24.

The RM for the shelter in make-believe Erskine Hall, introduced in the discussion in paragraph 4-2b(1), used the time-averaging method to compare the radiation safety of the seven rooms in the basement. The locations where the RM made the measurements are shown in Figure 4-6, page 4-25.

People were packed together in Room G, where the RM made the first and last readings. The choice of locations where readings were taken and the order in which they were taken was made before fallout arrived. Fallout arrived at Erskine Hall at 1009 hr*, and the first radiation

*Twenty-four hour time is used to prevent confusion between AM and PM. This time notation is used by airlines and the military services. The first two digits indicate the hour of the day, starting with zero at midnight, and the second two digits indicate the minutes after the hour. The 24-hr time in the afternoon is obtained by adding 12 to the 12-hr time in the afternoon (hours past noon). Thus, 1:10 PM (ten minutes past one) becomes 1310 hr, 2:20 PM becomes 1420 hr, etc. See Appendix B for a table to convert standard time to 24-hr time.

Table 4-2. An example of the use of the time-averaging method.

Location[a]		Measurements						
		Time		Survey Meter Reading (R/hr)			Time-Average Radiation Rate (R/hr) (total divided by 2)	Comments
Room Name	Location Number	Before	After	Before	After	Total		
G	1	1040	1054	0.41	0.74	(1.15)	0.575	Lowest rate
A	2	1041	1053	0.73	1.19	(1.92)	0.96	One-minute delay
B	3	1043	1051	0.69	0.95	(1.64)	0.82	
C	4	1044	1050	1.01	1.29	(2.30)	1.15	
F	5	1045	1049	1.32	1.55	(2.87)	1.435	
E	6	1046	1048	0.79	0.86	(1.65)	0.825	
D	7	1047		0.96			0.96	Central measurement

[a]/Figure 4-6 shows where these locations are in the basement of Erskine Hall.

NOTE: This example results from an imaginary situation at Erskine Hall in which the time-averaging method is used to compare the radiation safety of various rooms when radiation levels are rising rapidly. The numbers are presented here as they might be entered by the RM in the RM log. The location numbers are entered on the sheet before starting. The columns marked "Before" under both the "Time" and the "Survey Meter Readings" are filled in from top to bottom as the measurements are made, and then the columns marked "After" are filled in from bottom to top. The numbers in parentheses in the column marked "Total" are obtained by adding the "before" and "after" survey-meter readings at a location. The time-average radiation rate at a location, except for the central measurement, is obtained by dividing the total by two.

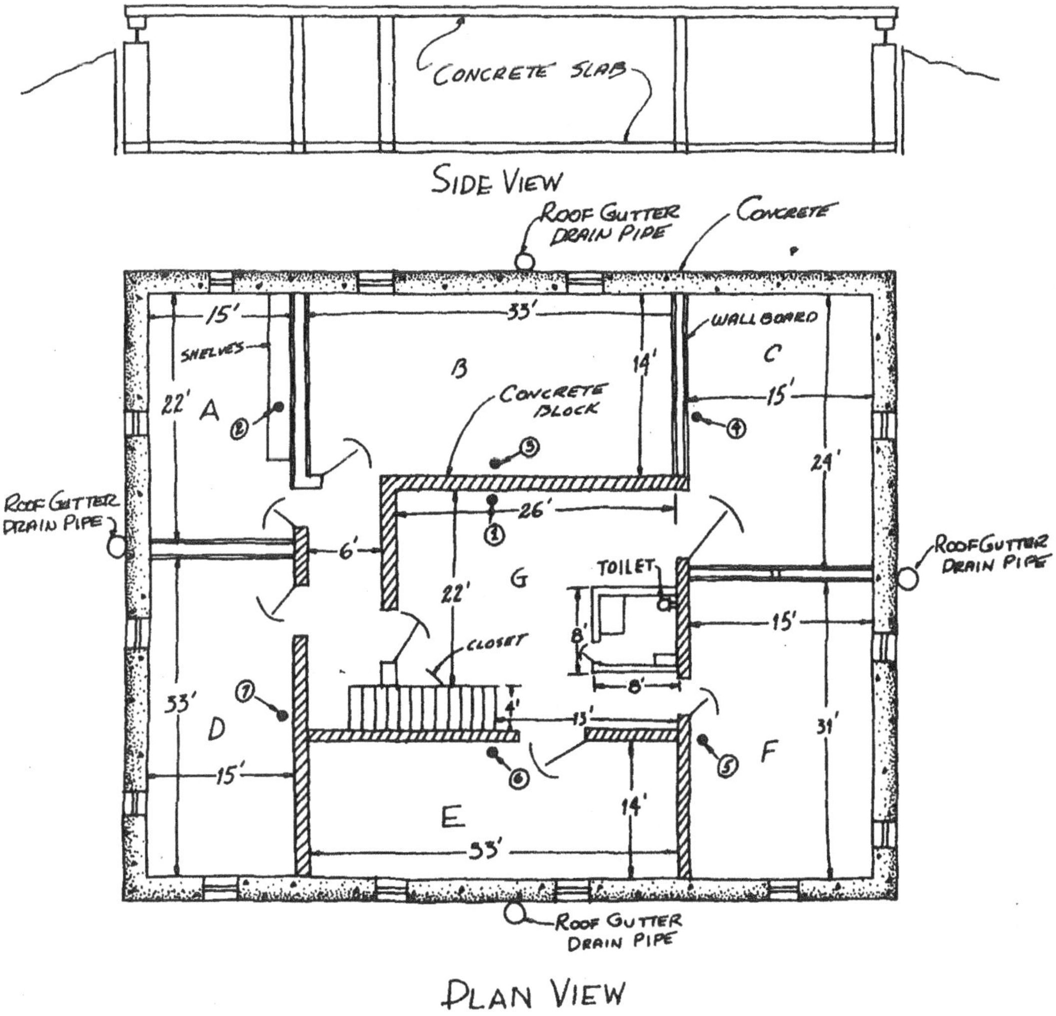

Figure 4-6. Locations of survey-meter readings for time-averaging in the basement of Erskine Hall are shown by dots and are identified by numbers in circles.

reading inside the shelter was made at location 1 at 1020 hr, as shown in Figure 4-5, page 4-17. A rapid survey throughout the basement roughly confirmed that Room G provided the best radiation protection. It was decided the first series of detailed measurements for time-averaging would begin at 1040 hr. The survey meter was brought to each designated location with enough time allowed to hold the meter in position for 10-15 seconds before the reading was taken. The first reading was taken at 1040 hr and the last at 1054 hr. Readings at location 1 were made seven minutes before and seven minutes after the central reading was taken at 1047 hr at location 7. Readings at location 2 were made six minutes before and six minutes after the central reading at location 7, and so on. While moving from location 2 to location 3, the RM was delayed by a disturbance between some occupants of the shelter, so the reading at location 8 was taken at 1043 hr instead of 1042 hr as initially planned. In order to maintain the same time interval between the "before" and "after" readings at locations 1 and 2, the "after" readings at those locations were delayed a minute to 1054 hr and 1053 hr, respectively, instead of 1053 hr and 1052 hr as initially planned. The two readings made at each location (except where the central reading was made) were added and divided by two to give an estimate of what the readings would have been at those locations at the same time the central reading was taken (1047 hr) at location 7. These time averages are listed at the bottom of Table 4-2, page 4-24. From these readings, it was confirmed that Room G (location 1) provided the best radiation protection in the basement of Erskine Hall. Note that the readings at locations 1 and 2 almost doubled between the "before" and "after" readings.

Another series of measurements for time-averaging should be made as soon as practical, within 20 minutes after the first series, to confirm the results of the first series of measurements.

If the first series of measurements for time-averaging shows that there is an unoccupied area of the shelter where the radiation levels are significantly lower, say 20 percent, than the area where the people are located, notify the Shelter Manager, and also inform him or her that you are going to make another series of measurements to check your results. The Shelter Manager may wish to double-check your results. If your second series of measurements confirms the results of your first series, then the Shelter Manager will need to consider the possibility of moving shelter occupants to this new location.

A number of factors should be taken into account before the decision is made to move or not to move. If the new location offers only a slight reduction (less than 20 percent) in radiation levels, a decision not to move may be made for several reasons, such as: (1) there may be less space, less desirable space, and/or not enough ventilation in the new location; 2) the location of the new space may result in higher radiation exposures to occupants while they walk to restrooms or to eating facilities; and 3) fire escape routes may not be as good. If the new location offers substantially lower radiation levels, a decision to move may be made in spite of such shortcomings, especially if it appears that the radiation intensity may climb to such high levels that the accumulated exposure may result in radiation sickness. Even if the current fallout is so light that radiation sickness is not likely, the Shelter Manager may decide that the occupants should move in order to be better prepared for the possibility of additional fallout from future attacks.

If a sudden squall or weather front with high winds and heavy rain strikes the shelter while you are in the process of taking readings for time-averaging, you may need to disregard your measurements and wait until the weather settles down before you try the readings again. You may not be able to tell whether a decrease in reading from one room to another results from the second room being safer or from a decrease in radiation level because fallout particles are temporarily being blown and washed away. The reading may change because of a combination of these two causes.

You should compare the radiation levels between the different areas at least every 12 hours, or whenever anything takes place that might move the fallout particles around, such as a heavy rain or windstorm. After the fallout has stopped coming down and the rates are not changing rapidly, it won't be necessary to use the time-averaging method for making these comparisons.

d. Finding and Covering Up "Leaks" in Gamma Shielding. After the safest locations have been found in the shelter and the people have moved there (if they weren't there already), use the survey meter to make detailed measurements of the radiation levels in and around the area where the people are located. During the first rapid, spread-out survey of the room, you may have noticed that your meter readings were higher in certain places within the room. This variance could be the result of uneven piling-up of fallout around and above the shelter; the layout of rooms, walls, and stairways; openings in the walls; and/or the use of lighter-weight construction materials in some places. It may be possible to use the survey meter to locate a specific place where gamma radiation is entering or "leaking" into the shelter to cause higher readings. When such an area is identified, any available materials should be used to cover it in order to reduce the level of radiation.

For the measurements you made to find the safest places in the shelter, you held the survey meter out from your body about two feet, and, in crowded rooms, people were asked to sit or lie down, so their bodies would cause less interference with the reading. But for finding gamma leaks, you can make use of that interference.

The survey meter responds to gamma rays almost equally as well from all directions. If gamma rays come in greater intensity from one particular direction, you can't detect the direction just by pointing the instrument toward it. But you may be able to use the shielding provided by your body and others to reduce the radiation coming from the direction where you and others are grouped together; the survey meter will then respond more to radiation coming from OTHER directions than from where you are standing. For example, if a group of people crowd around a survey meter and leave an opening in only one direction, the reading on the instrument will be caused mostly by radiation coming through the opening, providing there isn't a lot of radiation coming down through the ceiling or up from the floor. This method has not been tested in practice, and you may be able to improve it as you try it. Also, you may find that it does not work in your particular circumstances.

The measurements are made as follows: Select a starting place somewhere along a wall, at a corner, at a door or window in the shelter room. Mark that location on the floor or on the wall with a piece of tape or by writing directly on the surface. Use a letter to designate the room and a number to designate the place where the measurement is taken in the room. For example, the first measurement in Erskine Hall would be taken at a spot marked "G-1," because the room marked "G" on the shelter sketch is the room where the people are sheltered. Hold the survey meter against your waist and face the wall with the survey meter against the wall or a few inches from it. Have an assistant write down the location designation, the time, and the survey-meter reading in the RM log or on a sheet of paper.

Move three or four feet to your right or left (it doesn't matter which direction you go as long as you keep going in the same direction) along the wall and mark the location with the same letter as before, but with the number "2" ("G-2" in Erskine Hall, for example). Hold the survey meter as before, read the dial, and again record the location, time, and reading. Continue the measurements until you have gone completely around the room and have reached your starting point.

It is important that you take readings in the middle of doorways, windows, other openings or irregularities in construction. You may have to break your pattern of equal spacing between measuring locations in order to obtain these special measurements.

You will very likely be taking these measurements while fallout is still coming down. As you go around the room, the readings will become higher and higher in a fairly regular pattern unless you find a place that appears to be a "leaky" area. As you approach such a place, the readings will increase more between readings than before, and as you go beyond the area, there will not be as much of an increase in the reading; in fact, there may be a decrease in the reading. Because the radiation levels will be increasing at a fairly regular rate under most conditions, you should try to MAINTAIN AN EQUAL TIME INTERVAL BETWEEN MEASUREMENTS as you go around the room. A time interval of 20 or 30 seconds may be about right. Don't try to go too fast or you might not be able to keep up with the schedule. If you notice an area that appears to be "leaky," don't slow down. Continue with your measurement schedule around the room. You may need to ask the Shelter Manager to give you some assistance to make sure that nothing will interfere with your schedule of measurements.

After you have completed your measurements around the room, examine the numbers your assistant wrote down for indications of "leaky" areas. If you find any indication of such areas, tell the Shelter Manager. You should also tell him or her that you will need the assistance of several people to help you decide whether there is an actual leak of gamma radiation at the locations or whether the readings are a result of the way the scattered gamma radiation happens to be focusing at that location.

You will need to repeat your measurements in the vicinity of the suspected area, starting at the location just before the increased numbers were recorded, and make measurements, again AT REGULAR TIME

INTERVALS, until you have passed the suspected area; but this time the people in the vicinity of the area should be asked (possibly by the Shelter Manager, depending on the situation) to stand and press fairly close to you while you make each measurement. The shielding that is provided by their bodies will block out scattered gamma radiation that comes from different directions inside the room. If the readings still show an increase as you approach the area and a decrease as you go past it, there is a "leak" of gamma radiation in the area you are surveying. This leak could come from the area in front of you, or it could come from above (or below, if you are in an aboveground shelter). If the readings no longer show an increase as you approach the area and a decrease as you go past it, the previous reading (without the people standing closely behind you) was caused by the pattern of scattered gamma radiation in the room, not by a gamma leak.

If you are trying to find gamma leaks in an empty room, you may use the "front-to-back" method. In this method, your own body is used as a shield to try to find from what direction the gamma radiation is coming. Again, this method has not been tested in practice, and you may be able to improve it as you try it, or you may find that it won't work in your particular circumstances.

To try to find a gamma leak, hold the survey meter tightly against your stomach and face the area where you expect extra gamma radiation to be coming from. If you are working with the range-selector switch turned to "X0.1," wait a few seconds before taking a reading. This reading will be called a "front" reading. Turn around so your back faces the suspected leak, and with the survey meter still held tightly against your stomach, take another reading. This reading will be called a "back" reading. If there is more radiation coming from the direction you faced for the first reading than from the opposite direction, the front reading will be higher than the back reading. As you slowly turn around, you may notice that the meter needle goes through the lowest reading when you are facing a particular direction. The radiation leak is then at your back. Repeat these "front-to-back" readings at different places and directions until you have a fairly good idea of where the extra radiation is coming from. The difference between the front and back readings may be made greater, if the radiation is actually coming from one direction more than another, by having several others stand alongside and behind you when you make the measurements. The extra shielding provided by their bodies will take out more of the radiation from the rearward direction, which is what you want to do while making this type of measurement.

When you are fairly certain you have found a radiation leak, tell the Shelter Manager. A work party should be organized to build a gamma barrier to cover up the leak. If you had the time and opportunity, you should have gathered materials for this purpose before fallout arrived, as discussed in paragraph 4-2b(3). Work on construction of this barrier should begin as soon as possible, before the radiation climbs to higher levels. The barrier can be improvised from any materials on hand. If you have lumber, nails, and carpenter's tools available and have hauled piles of earth or sand into the shelter before fallout arrived, you may be able to construct a very good barrier. Stacks of bricks will also make good barriers. If these materials aren't available, items such as furniture, books, magazines, newspapers, and water containers may be used.

While the barrier is being constructed, do not forget to take the regular readings which tell whether the radiation levels are rising or falling. Write these readings on a piece of paper or on a form as shown in Figure 4-5, page 4-17, and tape or tack it to a wall or post near the place where the reading was made.

After the barrier is constructed, take several measurements of the kind you took to find the leak, to see if the radiation leak has been covered up. If you found the leak by taking a series of measurements from one side of the area to the other, with several people standing closely behind you, you should repeat that kind of measurement. You should be able to tell by these measurements if the barrier has improved the shielding in the leak area, or if more work is required on the barrier. If there is no change in these readings from your earlier readings, there is a possibility that the barrier may have missed the area through which the extra gamma radiation is passing. In this case, more work should be done to locate the leak and construct the barrier.

Again, let us look at Erskine Hall as an example. The shelter sketch is shown in Figure 4-6, page 4-25. In making a detailed survey of Room G, the RM found readings in two places which were 15-30 percent higher than at other places in Room G. One location was by the closet under the stairs and the other location was by the open door to Room F.

The reading by the stair closet was about 15 percent higher than elsewhere. The radiation was assumed to be coming from above, through the stairways. The Shelter Manager, RM, and Unit Leaders decided not to pile material on the stairs because the occupants would then have trouble getting out if there were a fire. Instead, they blocked off an area by the closet and planned to rotate people in and out of that area so the radiation dose would be evenly spread out among people in radiation sensitivity category Y/A, Table 4-1, page 4-4.

The reading by the door to Room F was about 30 percent higher than elsewhere. In the time-averaging readings, Room F (location 5) was found to have a higher reading than the other rooms, as shown in Table 4-2, page 4-24. This higher reading was expected, because in improving the radiation shielding of the shelter, all the windows around the basement had been covered except two in Room F. Materials were not available to construct baffles around these windows, such as shown in Figure 4-4, page 4-11. Instead, a wall of earth was piled up a few feet away from the window to shield the window against gamma radiation coming from fallout on the ground beyond the earth barrier. It was considered absolutely essential to leave these windows open to provide cooling for the people packed in Room G. Fresh air was coming in from those windows, passing through the open door to Room G, and flowing out the door by the stairs.

After examining the sketch of the floor plan, it was decided that a hole could be knocked in the wallboard partition to allow air to flow between Rooms C and F and the door between Rooms C and G could be left open. The door between Rooms G and F could then be closed and covered with a barrier.

The hole between Rooms C and F was made on the far side from the door by the outside wall, so the gamma rays from the two open corner windows would not have a direct open path to the door between Rooms C and G. The door between rooms F and G was closed, and a stack of bricks was built in front of it.

These measures reduced the radiation in Room G near the door to Room F to levels that were about the same as elsewhere in the room (except by the stairway closet). Ventilation became much better for the people along the north half of the room, but the people in the hall leading to Room F soon complained about lack of ventilation. The bricks in front of the door to Room F were restacked so there were one- to two-inch gaps between the bricks on the bottom four layers. The door was propped open a few inches so air could flow through the gaps left between the bricks. Another wall of bricks, only six layers high, was constructed about six inches back from the door-high stack of bricks, to block off gamma rays coming through the gaps.

e. Gamma Shielding by People. In Table 1-2, page 1-11, the human body is listed with a density of 0.4 relative to concrete. The shielding effect of human bodies can be used to provide extra protection. This protection would be of particular benefit to those people with the greatest sensitivity to radiation, namely, children and pregnant women. If the estimated or projected radiation exposures look as if they may become high enough to cause radiation sickness and other ways to decrease or avoid radiation exposure are not possible, this shielding method could be used. It would be expected that this extreme measure of providing shielding would be used only during the first 24 hours after fallout arrives, when the radiation hazard is by far the most severe.

Ordinarily, people in most shelters will be sitting or reclining on the floor most of the time. More gamma radiation will be blocked if the people are standing up, because their bodies will then absorb some of the gamma rays coming from the ceiling as well as those coming from the walls. This shielding, provided by people who are standing, could provide an extra measure of protection for children, mothers with infants, and pregnant women. By forming a two- or three-person-deep circle around the more radiation-sensitive occupants of a shelter, these individuals can possibly be spared high radiation exposures that would be especially detrimental to them. The survey meter should be used to find the arrangement of people that provides the best shielding.

Children and infants may be provided additional protection from overhead radiation by placing them underneath beds, desks, tables, or other suitable items. People with less radiation sensitivity may then sit or lie on top to provide additional shielding.

The RM may verify the shielding effect provided by people by reading the survey meter at different levels in the middle of a room full of people who are standing up. In basement shelters, where no gamma radiation comes up through the floor, the survey meter reading at the floor might be as much as ten times lower than the reading at waist height at the wall. The radiation may even be undetectable at the floor. In high-rise shelters where much of the gamma radiation comes in horizontally through the walls and some comes up at different angles through the floors, this effect won't be as dramatic.

f. Keeping Track of Everyone's Radiation Exposure (Group Dosimetry). The radiation hazard will be worst throughout the first 24 hours after each fallout cloud arrives. It is important to start keeping track of everyone's radiation exposure right away, as soon as fallout begins to arrive. In most shelters the radiation levels will be different as you move from one place to another. In these shelters each Unit Leader should have a dosimeter. The readings on the Unit Leader's dosimeter will be used to fill out the radiation exposure record of each member of the unit. For this reason, every member of the unit should stay close to the leader, especially during the first 24 hours after fallout arrives. This method of estimating individual exposures is called GROUP DOSIMETRY.

If any member of the unit needs to make an urgent trip to some area where the radiation level is higher and for a length of time such that the person's radiation exposure might be a few roentgens higher than the rest of the unit, special arrangements should be made. The Shelter Manager and RM should be consulted if the trip is unusual. An extra entry should be made on the individual's radiation exposure record for such trips.

Trips to restrooms and drinking fountains in areas of higher radiation levels should be limited in number and length. The Unit Leader should make about the same number of trips as other unit members at about the same times for the same length of time. The dosimeter should be worn by the Unit Leader on these trips to get an idea of how much exposure is received during these trips. If some members need to make additional trips, the extra exposure should be estimated by the Unit Leader, with help from the RM if necessary, and entered on the members' radiation exposure records.

You, the RM, should very carefully monitor your own exposure and make forecasts on your future exposures so you will not exceed the limit of exposure set in Row A of the Penalty Table (Table 4-3, page 4-38). Your experience and training make you very valuable to the occupants of the shelter.

A dosimeter hung on the wall or a post at eye level or higher will show a higher radiation exposure than a dosimeter carried on a person in the same area. The person's body shields the dosimeter from some of the gamma radiation. If the person wearing the dosimeter is surrounded by many people who are standing up, the reading on that person's dosimeter will be even lower because of the gamma shielding provided by the people's bodies.

During the first 24 hours after fallout begins to come down, entries should be made every 4 hours in each person's radiation exposure record. The Unit Leader should check each entry on each record kept in his unit. The RM should spot-check records throughout the shelter and look for entries which seem too high or too low. Such entries may be due to faulty instruments or to shielding conditions which the RM should know about. It is important that these situations be corrected as soon as possible.

Sample radiation exposure records from Erskine Hall are shown in Figures 4-7 and 4-8. The radiation exposure record in Figure 4-7 shows what a dosimeter would read if it were mounted at location 1, where survey meter readings were taken for Figure 4-5, page 4-17. The radiation exposure record taken from dosimeters clipped to the clothing of adults on the edges of Room G would have entries which may be less than 75 percent of the entries in Figure 4-7, due to the shielding effect of their own bodies and others. The entries on records of those in the interior of the room would be even lower.

In Figure 4-8 the radiation exposure record is shown for John Doe, an infant. His radiation sensitivity category is "CHILD," as listed in Table 4-1, page 4-4. This record was maintained by his father, James Doe, who was made the leader of the shelter unit in which the Doe family was placed. The radiation levels in Erskine Hall started to climb a second time at 1645 hr on July 5, 1989, as shown by the survey-meter readings in Figure 4-5 page 4-17, indicating the arrival of another cloud of fallout. By 1745 hr the radiation level had reached 5 R/hr at location 1 and was still climbing. It was decided that human body shielding would be used to protect those in the first two radiation sensitivity categories. This special shielding, involving all the people in the shelter, began at 1800 hr, as shown on the radiation exposure records in Figure 4-8, and reduced John Doe's exposure to less than half of what it would have been without this special shielding. On the second day, 24 hours after fallout arrived, special shielding was terminated, but partial shielding for John Doe was provided by the members of his shelter unit. The next 13 entries were made on a daily basis instead of every four hours. On July 18, the occupants of Erskine Hall were relocated to a shelter in an area with much lighter fallout.

4-5. <u>After Fallout Has Stopped Coming Down</u>.

a. <u>Forecasting Radiation Exposure</u>. When the survey meter readings level off and then continue to decrease, the arrival of fallout from that particular cloud at your location has almost ended. If no more fallout clouds arrive, the radiation levels will continue to decrease rapidly.

RADIATION EXPOSURE RECORD

Name LOCATION 1

Home Address

Social Security No.

Shelter Address ERSKINE HALL

Name of Shelter Unit Leader

Rad. Sensitivity Category

Hour and Date	Added Exposure (R)	Total Exposure To Date	Comments
1400 5-7-89	8	8	FALLOUT BEGAN AT 1009
1800 5-7-89	14	22	
2200 5-7-89	23	45	
0200 6-7-89	15	60	
0600 6-7-89	12	72	
1000 6-7-89	9	81	END OF FIRST 24 HR
1000 7-7-89	35	116	
1000 8-7-89	19	135	
1000 9-7-89	14	149	
1000 10-7-89	10	159	
1000 11-7-89	8	167	
1000 12-7-89	7	174	END OF FIRST WEEK
1000 13-7-89	6	180	
1000 14-7-89	5	185	
1000 15-7-89	4	189	

FRONT SIDE

Hour and Date	Added Exposure (R)	Total Exposure To Date	Comments
1000 16-7-89	4	193	
1000 17-7-89	3	196	
1000 18-7-89	3	199	SHELTER VACATED
1000 19-7-89	3	202	END OF 2nd WEEK
1000 22-7-89	7	209	
1000 26-7-89	9	218	END OF 3rd WEEK
1000 2-8-89	8	226	END OF FIRST MONTH

BACK SIDE

Figure 4-7. Sample dosimeter readings at location 1 in make-believe Erskine Hall.

RADIATION EXPOSURE RECORD

Name JOHN DOE

Home Address SOMEWHERE, USA

Social Security No. NONE--INFANT

Shelter Address ERSKINE HALL

SOMEWHERE, USA

Name of Shelter Unit Leader JAMES DOE

Rad. Sensitivity Category CHILD

Hour and Date	Added Exposure (R)	Total Exposure To Date	Comments
1400 5-7-89	6	6	FALLOUT BEGAN 1009
1800 5-7-89	10	16	SPECIAL SHIELDING BEGUN
2200 5-7-89	7	23	
0200 6-7-89	5	28	
0600 6-7-89	4	32	
1000 6-7-89	3	35	UNIT SHIELDING BEGUN
1000 7-7-89	18	53	
1000 8-7-89	10	63	
1000 9-7-89	7	70	
1000 10-7-89	5	75	
1000 11-7-89	4	79	
1000 12-7-89	3	82	END OF FIRST WEEK
1000 13-7-89	3	85	
1000 14-7-89	3	88	
1000 15-7-89	2	90	

FRONT SIDE

Hour and Date	Added Exposure (R)	Total Exposure To Date	Comments
1000 16-7-89	2	92	
1000 17-7-89	2	94	
1000 18-7-89	2	96	EVACUATED
1200 18-7-89	6	102	TRIP TO SAFERVILLE
1000 19-7-89	—	102	END OF 2nd WEEK
1000 26-7-89	1	103	END OF 3rd WEEK
1000 2-8-89	—	103	END OF FIRST MONTH

BACK SIDE

Figure 4-8. Sample radiation exposure record for the fictitious John Doe, as filled out by his father. The estimated effect of human shielding may be seen by domparing these entries with the readings of an exposed dosimeter shown in Figure 4-7.

The highest radiation exposure at a given place in a shelter will accumulate during the first 24 hours after fallout arrives. After these first 24 hours have passed, there are two general rules which can be used to forecast the radiation exposure, as follows:

RULE 1: The radiation exposure at a given place during the entire WEEK following the arrival of fallout is unlikely to be more than 2-1/2 TIMES the exposure during the first 24 hours.

RULE 2: The radiation exposure at a given place during the entire MONTH following the arrival of fallout is unlikely to be more than 3-3/4 TIMES the exposure during the first 24 hours.

If the fallout comes from distant ground bursts and doesn't arrive at your shelter until 24 hours or more after the explosions, the numbers in Rules 1 and 2 may be slightly greater. For example, if the fallout takes about 36 hours to get to your shelter, the number 2-1/2 in Rule 1 will be increased to 3.0 and the number 3-3/4 in Rule 2 will be increased to 4.5.

If the fallout takes about 48 hours to get to your shelter, the corresponding numbers will be increased to about 3-1/3 and 5-1/3, respectively. When the fallout takes a long time to arrive, the radioactivity will have decayed a great deal. If the fallout comes from a large number of ground bursts of large-yield weapons, as might take place on military targets, the fallout may still be hazardous even though it may take 48 hours to arrive at your shelter.

If the fallout comes from closer ground bursts and arrives at your shelter in 12 hours or less after the explosions, the numbers in Rules 1 and 2 will be less. More than half of the total exposure in a week will accumulate in the first 24 hours after fallout arrives. The number 2-1/2 in Rule 1 will be decreased to between 1-1/2 and 1-3/4, and the number 3-3/4 in Rule 2 will be decreased to between 1-3/4 and 2-1/2.

Exposure forecasts can be made using the seven-ten rule described in paragraph 1-10d when all the fallout is the same age, when the time of the explosion is known fairly well, and when there are no weathering effects. These circumstances are unlikely in a modern, full-scale nuclear war. The Radiological Defense Officer in the local Emergency Operating Center (EOC) may be able to provide further guidance on estimated radiation exposure.

The general rules given above can be used to make forecasts for the possibility of radiation sickness among a group of people in a given shelter. If the radiation exposure of an average adult is 60 R or less at the end of 24 hours after fallout arrives and that person remains in the same place, that person's accumulated radiation exposures will be expected to be less than 150 R in one week and less than 225 R in one month, providing no additional fallout arrives. According to the Penalty Table, page 4-38, that person should require no medical care in the first week, but the exposure in a month would exceed the limits set in the Penalty Table for not requiring medical care.

If it appears that the radiation exposure of average adults will be more than a preselected value, such as 60 R, at the end of the first 24 hours after fallout arrives at the shelter, the local EOC should be notified. Some emergency action may be possible which will reduce the accumulated radiation exposure and thus prevent radiation sickness among these people.

Again, let us look at the made-up example provided by Erskine Hall. The radiation exposure record for a dosimeter mounted at location 1 is shown in Figure 4-7, page 4-34, and the survey-meter readings for that location are shown in Figure 4-5, page 4-17. The first detection of fallout was made outside the shelter at 1009 hr on July 5. It was estimated that this fallout resulted from many large-yield ground bursts on military targets about 250 km (150 miles) upwind during the night before, at around 2100 hr on July 4. The radiation level from this fallout reached a maximum value at around 1330 hr on July 5, indicating that most of the fallout destined for Erskine Hall from these explosions had reached the ground by this time. The fallout took 13 hours to reach Erskine Hall. It kept coming down for about 3-1/2 hours.

A distant explosion was heard at 1500 hr on July 5, in the direction of a city located about 50 km (30 miles) upwind. The fallout from this explosion began to arrive at Erskine Hall at about 1645 hr, an hour and 45 minutes after the explosion was heard. This fallout was more radioactive than the older fallout from the distant explosions. Being fresher, it would decay faster. This fallout kept coming down for about 2-1/2 hours and added to the radiation levels which were already there from the older fallout.

At the end of the first 24 hours after fallout arrived, at 1000 hr on July 6, the accumulated radiation exposure by the dosimeter at location 1 was 81 R, as shown in Figure 4-7, page 4-34. After one week, the accumulated radiation exposure was 174 R, 2.15 times the exposure during the first 24 hours. After one month, it was 226 R, 2.79 times the exposure during the first 24 hours.

b. The Penalty Table. An adult will not normally need medical care when the whole body is exposed to the quantities of radiation listed in Row A of Table 4-3 if the exposure is spread out over the listed periods of time. Rows B and C are intended to be used for making decisions on performing urgent missions which may involve the risk of increased radiation exposure.

Each person can tolerate a certain amount of sunshine on bare skin in an afternoon without getting a painful sunburn. Similarly, each person can be exposed to a certain amount of whole-body gamma radiation within a certain period of time without getting sick. The Penalty Table (see Table 4-3) shows in row A what exposures might be received by an average adult without requiring medical care, when the exposure is spread out over different periods of time. Infants, small children, and pregnant women should be given special consideration when possible, because they are more likely to have radiation sickness at lower levels of radiation exposure than other individuals of the general population.

Table 4-3. The penalty table[a]

Medical care will be needed by--	Accumulated radiation exposure (R) in any period of		
	a	b	c
	One week	One month	Four months
A NONE	150	200	300
B SOME (5 percent may die)	250	350	500
C MOST (50 percent may die)	450	600	---

[a]This table is taken from Radiological Factors Affecting Decision-Making in a Nuclear Attack, National Council on Radiation Protection and Measurements, Report No. 42.

For most shelter occupants, the exposures in row A should not be exceeded. If the radiation levels reach 10 R/hr in the shelter and continue to climb, it is possible that the accumulated exposure in one week will be greater than 150 R. In this case, the local EOC should be notified. Some emergency action may be possible which will reduce the accumulated radiation exposure and thus prevent radiation sickness in the shelter.

c. Use of the Penalty Table as a Guide for Operations. The Penalty Table was developed to provide a simple guide when decisions must be made that will involve some risk. The choice of the numbers was based on judgment derived from extensive clinical radiotherapy experience, pathological studies of radiation-accident victims, and laboratory experience with numerous large and small animals. There is no directly-applicable disaster or laboratory experience involving humans that clearly supports the choice of all of the numbers in the Penalty Table. There is also no satisfactory biological model or mathematical formula relating radiation effects (of the type considered here) to exposure rates and durations that provides a satisfactory basis for deriving the amounts of exposure indicated in the table for time periods greater than one day. These are the best numbers available at the present time for this purpose.

Three examples of the Penalty Table are given here:

Example 1. It would be best if everyone's radiation exposure could be kept as low as possible, but due to wartime conditions, some individuals may have to spend some time in areas of higher radiation levels. Suppose you are trying to limit their radiation exposures to levels resulting in low risk. The numbers in Row A of Table 4-3 apply in this case. According to these numbers, it would be necessary to limit the total radiation exposure of individuals to less than 150 R in any one week (column a), 200 R in any one month (column b), and 300 R in any four-month period (column c).

For example, if individuals were exposed to the one-week limit of 150 R (column a) within the first week, then the limit for additional exposure during the following three weeks of the first month, to keep within the one-month limit (column b) would be 200 R - 150 R = 50 R. This additional exposure of 50 R could be received at any rate (for example, by going outside the shelter into areas of higher radioactivity) during the following three weeks of the first month, without exceeding the one-week or one-month limits in the Penalty Table. However, if this additional exposure of 50 R were received, for example, within the second week, then the individuals would have to be kept completely free of further exposure (which may not be possible) during the remainder of the first month to keep within the one-month limit for Row A (200 R). Similarly, if the individuals were exposed to the limit of 200 R in the first month, without exceeding 150 R in any one week of that month, the limit of additional exposure for the following three months of the first four months (column c) would be 100 R, for a total of 300 R (200 R + 100 R) in four months.

Example 2. Suppose you need to conduct operations at the intermediate level of radiation exposure, involving significant medical risk (Row B), justified by highly critical emergency situations. The decision to conduct such operations must involve the Shelter Manager.

In this case, the decision-maker may find it necessary to allow greater exposure than one or another of the limits indicated in Row A but would be constrained whenever possible by other limits in Row A and always by limits in Row B of the Penalty Table, page 4-38.

For example, if individuals who have been exposed to 150 R within the first week are required in some emergency to be exposed to an additional 200 R during the remainder of the first month (for a total of 350 R in the first month), it is desirable, if possible, that the one-week constraint for Row A (column a) be observed by allowing no more than 150 R of this additional exposure during any one week within that month, even though the one-month limit (200 R) and four-month limit (300 R) for Row A will have been exceeded and the one-month limit (350 R) for Row B will have been reached. If it is not possible to keep within any of the constraints for Row A, then the Row B constraints have to be applied. In other words, you try to keep exposure in any one week as far as possible below 250 R and to limit the exposure during the first month to 350 R. Any additional exposure after this first month must be kept as far as possible below the additional 150 R which would attain the four-month limit of 500 R (Row B).

As in example 1, the decision-maker could schedule exposures in a variety of ways within the constraining limits to meet the work required by the problem at hand.

Example 3. Suppose you need to conduct operations at the high levels of medical risk (Row C), justified only by extremely critical emergency situations. Again the decision to conduct such operations must involve the Shelter Manager. Those activities that could result in saving a significant number of lives may call for the deliberate exposure of some persons at the highest constraint levels, where radiation sickness and a 50 percent probability of death are expected (Row C). If such situations arise, the decision-makers would use for guidance Row C of the table in a manner similar to that discussed for the low- and intermediate-risk rows (A and B) in examples 1 and 2.

After a time of no more than two weeks, it should be possible to move people from areas of high radiation levels to areas of lower radiation levels. In the areas of lower radiation levels, people should be able to get outside and work for different lengths of time as long as their radiation exposures stay within the limits of Row A of the Penalty Table. The "one-month" and "four-month" columns of the Penalty Table are intended primarily for these situations. No one should have to stay totally confined inside the shelters for more than two weeks, although people may have to live in them in some locations for longer periods.

d. Checking Radiation Levels Beyond the Immediate Shelter Area. Sometime no later than 24-30 hours after fallout has begun to come down, you (the RM) should take the survey meter and check the radiation levels in rooms next to the shelter area and on the way to the outside. The purpose of this exploration is to get an idea how dangerous the levels are outside the immediate shelter area, to estimate the risks in emergency operations, and to forecast when people could leave the shelter for short periods and when they could move to safer areas if needed.

Your experience and training make you very valuable to the occupants of the shelter. You should very carefully monitor your own exposure and make forecasts on future exposures so you will not exceed the limit of exposure set in Row A of the Penalty Table, page 4-38.

If you used the time-averaging method to find the safest location in the shelter and the fallout pattern hasn't been shifted by wind or rain, you may use the results of those measurements to estimate the radiation levels in the other rooms which you checked, by using the RATIO METHOD. Suppose you stayed near location 1, your "home base," during the first 24 hours after fallout arrived. Now you want to find out how high the radiation level is at location 5. Suppose you included location 5 in your time-average comparison. Then you can estimate the present reading at location 5 by first finding the ratio of the time-average reading at location 5 to the time-average reading at location 1. Then multiply this ratio times the current reading at location 1 to get the current reading at location 5. In other words,

$$\text{Current reading at location \#5} = \frac{(\text{Time-average reading at location \#5})}{(\text{Time-average reading at location \#1})} \times \text{Current reading at location \#1}$$

The measurements at Erskine Hall will be used as an example. Suppose we would like to know what the survey-meter reading would be at location 5 in Erskine Hall at 2000 hr on July 5, without actually taking the survey meter to the location. We have been making measurements regularly at location 1, as shown in Figure 4-7, page 4-34. We have the set of time-average measurements that were made earlier for seven locations, including locations 1 and 5, as listed in Table 4-2, page 4-24. To get the current reading (at 2000 hr) at location 5 without taking a survey meter to that location, the following steps are taken:

(1) the current reading (at 2000 hr) at location 1 is found to be 6.0 R/hr.

(2) the time-average reading at location 5 was 1.435 R/hr.

(3) the time-average reading at location 1 was 0.575 R/hr.

(4) the ratio of the time-average reading at location 5 to the time-average reading at location 1 is 1.435/0.575 = 2.5.

(5) the current reading at location 5 is estimated by multiplying the ratio obtained in step (4) times the current reading at location 1, which yields 2.5 x 6.0 = 15 R/hr.

If more than one set of time-averaging measurements has been made, be sure the calculate the ratio with readings that were made in the same set of measurements.

Once the ratio of the time-average readings has been calculated, that same ratio can be used to estimate the reading at the remote location at any other time, assuming that the fallout pattern hasn't been shifted by rain or wind. For example, the estimated reading at location 5 in Erskine Hall at 2100 hr would be 2.5 times the reading at location 1 at that time, which is 5.5. The estimated reading at location 5 at 2100 hr would be 2.5 x 5.5 = 13.75 R/hr.

You may use the ratio method to estimate the radiation levels; first, at various strategic locations inside your shelter building and, later, at various locations outside your building. First take a reading at your home-base location. Then take the survey meter (wear a dosimeter) to the strategic location and take a reading there. You will not need to use the time-average method after 24 hours after the last particles of fallout have arrived because the radiation levels will be decreasing slower than 1 percent per minute. The ratio of the reading at the strategic location to the reading at the home base can be used to estimate readings at the strategic location by multiplying that ratio times the home-base readings.

As an example, the RM at Erskine Hall measured 2.1 R/hr at location 1, the home base, at 1000 hr on July 6. The RM took the survey meter up the stairs and made a quick trip into the lobby of Erskine Hall, where the survey-meter reading was 85 R/hr. The ratio of the lobby to home-base reading was 40. By 1000 hr on July 7, the home-base reading was 1 R/hr. The ratio of 40 was used to estimate that the radiation level in the lobby at that time was 40 R/hr.

At that time the RM took the survey meter upstairs and out to the street in front of Erskine Hall, where he measured a radiation level of 105 R/hr. His dosimeter showed an increase of 2 R for this trip, which he made as quickly as possible. The street to home-base ratio of readings was thus determined to be 105.

e. Leaving the Shelter. When the exposure rates outside the shelter are known, Table 4-4 may be used as a general guide for permissible activities. Decisions on how much exposure may be allowed should be based strictly on the Penalty Table (page 4-38). Unit Leaders should continue to keep close track of the radiation exposure of each member until shelter is no longer required. If the shelter is vacated and people are moved to other shelters, it would be preferable if units remained together. Exposure records must go with the individuals to whom they belong.

If the fallout is relatively young (two or three hours since fallout stopped coming down) and the radiation levels are decaying rapidly, greater relaxation of shelter control can be tolerated than indicated in Table 4-4. Conversely, if the fallout is relatively old (several days or weeks), more rigid control would be required.

Table 4-4. General guide for permissible activities outside the shelter

If the outside exposure rate (R/hr) is	Permissible activities
More than 100	Outdoor activity may result in sickness or death. Occasions which might call for outside activity are (1) risk of death or serious injury in present shelter from fire, collapse, thirst, etc., and (2) present shelter is greatly inadequate--might result in fatalities--and better shelter, available for occupancy, is known to be only a few minutes away.
10-100	Time outside of the shelter should be held to a few minutes and limited to those few activities that cannot be postponed. All people should remain in the best available shelter no matter how uncomfortable.
2-10	Periods of less than an hour per day of outdoor activity are acceptable for the most essential purposes. Shelter occupants should rotate outdoor tasks to distribute exposures. Outdoor activities of children should be limited to no more than 10 to 15 minutes per day. Activities such as repair or exercise may take place in less than optimum shelter.
0.5-2	Outdoor activity (up to a few hours per day) is acceptable for essential purposes such as fire fighting, police action, rescue, repair, securing necessary food, water, medicine, and blankets, important communication, disposal of waste, exercise, and obtaining fresh air. Eat, sleep, and carry on all other activities in the best available shelter.
Less than 0.5	No special precautions are necessary for operational activities. Keep fallout from contaminating people. Sleep in the shelter. Always avoid unnecessary exposure to radiation.

Appendix A

GLOSSARY

The meanings of some of the specialized words and abbreviations used in this handbook are provided in this list, which is arranged alphabetically.

absorbed dose - the energy imparted to matter by ionizing radiation per unit mass of irradiated material at the point of interest. The unit of absorbed dose is the rad.

air burst - the explosion of a nuclear weapon at such a height that the fireball does not touch the earth's surface. Fallout from an air burst is negligible.

alpha particle - a positively charged nuclear particle identical with the nucleus of a helium atom that consists of two protons and two neutrons and is ejected at high speed from the nucleus of certain atoms in radioactive decay processes.

alpha radiation - rays of alpha particles.

alpha ray - an alpha particle moving at high speed, or a stream of such particles.

atom - the smallest particle of an element that still retains the characteristics of that element. Every atom consists of a positively charged central nucleus, which carries nearly all the mass of the atom, surrounded by a number of negatively charged electrons, so that the whole system is normally electrically neutral.

background radiation - nuclear radiations arising from within the body and from the surroundings to which individuals are always exposed. The main sources of the natural background radiation are potassium-40 in the body, potassium-40, thorium, uranium, and their decay products present in rocks and soil, and cosmic rays.

beta burn - damage to the skin caused by prolonged contact with particles that emit beta radiation.

beta particle - an electron (negatively charged particle) or a positron (positively charged particle) ejected at high speed from the nucleus of certain atoms in radioactive decay processes.

beta ray - a beta particle moving at high speed, or a stream of such particles.

beta radiation - rays of beta particles.

blast wave - a violent pulse of air in which the pressure increases sharply at the front, accompanied by winds, propagated from an explosion.

bone seeker - any compound or ion that migrates in the body preferentially into bone.

contamination - the deposit of radioactive materials on the surfaces of structures, areas, objects, or personnel.

cumulative dose - the total dose resulting from continued or repeated exposures to radiation.

curie (abbr., Ci) - unit of radioactivity equal to 3.7×10^{10} disintegrations per second.

decontamination - the removal of radioactive material from a structure, area, object, or person, or the reduction of radiation from a surface or area by covering it.

dose - a general term indicating the quantity of radiation or energy absorbed.

dose equivalent (abbr., DE) - a quantity that is related to the expected detriment resulting from exposure to any kind of nuclear radiation, defined as the product of the absorbed dose in rads and modifying factors; the unit of DE is the rem.

dose rate - absorbed dose delivered per unit time.

dosimeter - an instrument for measuring accumulated exposure to nuclear radiation.

dosimetry - the theory and application of the principles and techniques involved in the measurement and recording of radiation doses and dose rates. Its practical aspect is concerned with the use of various types of radiation instruments with which measurements are made.

electron - an elementary particle having a negative electric charge of 1.6×10^{-19} coulomb and a rest mass 1/1836 that of the proton. In atoms, electrons surround the positively charged nucleus.

element - one of the known chemical substances that cannot be divided into simpler substances by chemical means.

emergency services - elements of government that are responsible for the protection of life and property, such as fire, police, welfare, and rescue services.

<u>EOC (Emergency Operating Center)</u> - a well-protected headquarters at various levels of government, such as city, county, state, or region, with two-way radio and telephone communications with shelters, emergency services, other EOCs, and various government headquarters.

<u>exposure</u> - a quantitative measure of gamma or x-ray radiation at a certain place, based on its ability to produce ionization in air, measured in units of roentgens.

<u>fallout</u> - the process of the settling to the earth's surface of airborne particles containing radioactive material following a nuclear explosion; also refers to the particles themselves. Early fallout, also called local fallout, is that fallout which settles to the surface of the earth during the first 24 hours after a nuclear explosion. Delayed fallout, also called worldwide fallout, is that fallout which settles to the surface of the earth at some time later than the first 24 hours after a nuclear explosion. Most of the fallout from a surface burst will be deposited within 24 hours after a nuclear explosion and within 400 to 500 miles downwind from the explosion.

<u>fallout half-value thickness</u> - the thickness of a given material which will absorb half the gamma radiation incident upon it. This thickness depends on the nature of the material--it is roughly inversely proportional to its density--and also to the energy of the gamma rays. These factors are specially calculated for fallout radiation and include all processes of attenuation of radiation.

<u>fallout protection factor (FPF)</u> - an indication of the degree of protection provided by a location against gamma radiation from fallout. The FPF for a location is defined as the ratio of the radiation exposure rate at 3 feet above a flat, smooth, large, open area to the radiation exposure rate at the location in question, when the same amount of fallout is deposited uniformly over both locations. If the FPF of a location is <u>one</u>, that location provides no protection against gamma radiation. This factor is also called the protection factor (PF). It is called "fallout protection factor" in this handbook because "protection factor" can mislead people into thinking that a location with a high "protection factor" will also protect against blast and thermal radiation.

<u>fallout shelter</u> - an enclosed area or place which can provide refuge and protection against fallout radiation by absorbing some or most of the radiation directed toward the shelter.

<u>fireball</u> - the luminous sphere of hot gases which forms a few millionths of a second after a nuclear explosion as a result of the absorption by the surrounding air of the radiation emitted by the extremely hot weapon residues. The exterior of the fireball is initially sharply defined by the luminous shock front and later by the hot gases themselves and may be visible for several minutes.

fission fraction - the fraction (or percentage) of the total yield of a nuclear weapon which is due to fission, the remaining fraction of the yield being due to fusion. For thermonuclear weapons the average value of the fission fraction is about 50 percent.

fission, nuclear - a nuclear transformation characterized by the splitting of a high-mass nucleus into at least two other nuclei of lower mass and the conversion of some of the initial mass into a relatively large amount of energy.

fission products - a general term for the complex mixture of substances produced as a result of nuclear fission. About 80 different fission fragments result from approximately 40 different modes of fission. The fission fragments, being radioactive, immediately begin to decay, forming additional (daughter) products, with the result that the complex mixture of fission products so formed contains over 300 different isotopes of 36 elements.

FPF - see fallout protection factor.

fusion, nuclear - a nuclear transformation characterized by the uniting together of two or more low-mass nuclei into a nucleus of higher mass and the conversion of some of the initial mass into a relatively large amount of energy.

gamma radiation - rays of high-energy photons from radioactive material.

gamma ray - a photon of high energy, or a stream of such photons, emitted by the nuclei of certain atoms in radioactive decay processes.

ground burst - a nuclear detonation at the surface of the earth, or at such a height above the earth that the fireball makes contact with the surface.

ground zero - the point on the surface of the earth vertically below, at, or above the point at which a nuclear explosion is initiated.

group dosimetry - a method for estimating radiation exposures of individual members of a group when there aren't enough dosimeters for each member to have one.

half-life (radioactive half-life) - the time in which half the atoms of a particular substance undergo radioactive decay.

high-risk areas - geographical areas in the United States estimated to be subject to a 50 percent or greater probability of receiving blast overpressures of 2 psi or more in a nuclear war, or to a 50 percent or greater probability of receiving a radiation exposure of 10,000 R or more.

hot spot - a localized surface area of higher than average radiation.

initial nuclear radiation - nuclear radiation (essentially neutrons and gammas) emitted from the fireball and the cloud column during the first minute after a nuclear explosion. The time limit of one minute is set somewhat arbitrarily as that required for the source of the nuclear radiations to attain such a height that only insignificant amounts of radiation reach the earth's surface.

ion - an atom or molecule that has lost or gained one or more electrons to become electrically charged.

ionization - the process of adding electrons to or removing electrons from atoms or molecules.

isotopes - forms of the same element having identical chemical properties but differing in their atomic masses due to different numbers of neutrons in their respective nuclei and also differing in their nuclear properties, such as half-life, energy, and type of nuclear radiation emitted.

kiloton energy - approximately the amount of energy that would be released by the explosion of 1,000 tons of TNT, defined precisely as 10^{12} calories, or 4.19×10^{19} ergs.

latency or latent period - the period of time between exposure to radiation and the detection of a specified effect of that exposure; or, for acute radiation sickness, the time during which no symptoms appear between the first reaction to radiation exposure and the later radiation sickness.

lethal radiation dose - the total-body radiation exposure required to cause death in 100 percent of a large group of people within a specified time period. For example, $LD_{100/60}$ indicates a dose which is lethal to 100 percent of the people exposed within 60 days after the exposure.

megaton energy - approximately the amount of energy that would be released by the explosion of one million tons of TNT, defined precisely as 10^{15} calories, or 4.19×10^{22} ergs.

midlethal or median lethal radiation dose - the short-term, total-body radiation exposure to cause death in 50 percent of a large group of people within a specified time period. For example, $LD_{50/60}$ indicates a dose which is lethal to 50 percent of the people exposed within 60 days after the exposure.

milliroentgen (mR) - 1/1000 of a roentgen. 1000 milliroentgens equal one roentgen.

neutron - an elementary particle having no electric charge and a rest mass of 1.675×10^{-27} kilogram. The neutron is a constituent of the nucleus of every atom heavier than ordinary hydrogen.

nuclear radiation - particulate and electromagnetic radiation emitted from atomic nuclei in various nuclear processes. The important nuclear radiations, from the weapons standpoint, are alpha and beta particles, gamma rays, and neutrons. All nuclear radiations are ionizing radiations, but the reverse is not true; X rays and nearly all ultraviolet radiation, for example, are included among ionizing radiations, but they are not nuclear radiations since they do not originate from atomic nuclei.

nuclear weapon - any weapon which attains its energy release from the fission or fusion of atomic nuclei.

nucleus - the positively charged central portion of an atom, composed of protons and neutrons and containing almost all of the mass of an atom but only a tiny part of its volume.

overpressure - the transient pressure, usually expressed in pounds per square inch, exceeding the ambient pressure, in the shock (or blast) wave from an explosion. The variation of the overpressure with time depends on the yield of the explosion, the distance from the point of burst, and the medium, whether air, water, or soil, in which the weapon is detonated. The peak overpressure is the maximum value of the overpressure at a given location and is generally experienced at the instant the shock (or blast) wave reaches that location.

PF - see protection factor or fallout protection factor.

photon - a packet of electromagnetic energy having zero mass and no electric charge. Visible light is made up of low-energy photons, and gamma rays are high-energy photons.

protection factor (PF) - this factor is called "fallout protection factor" in this handbook and is defined under that name. The term "protection factor" can mislead people into thinking that a shelter with a high protection factor will provide protection against blast.

proton - an elementary particle having a positive electric charge numerically equal to that of the electron and a mass of 1.672×10^{-27} kilogram. The proton constitutes the nucleus of the hydrogen atom and is a part of the nucleus of every atom.

rad - a special unit of absorbed dose equal to 100 ergs of energy imparted by ionizing radiation per gram of absorbing material, such as body tissue. The exposure rate measured at a point in roentgens /hr may be taken to be numerically equal to the absorbed dose rate in rad/hr at that point for external sources of gamma radiation.

radioactive decay - a spontaneous nuclear transformation in which a nucleus emits alpha or beta particles, often accompanied by gamma radiation, resulting in a progressive decrease in the number of radioactive atoms in a substance.

<u>radioactivity</u> - the spontaneous emission of radiation, generally alpha or beta particles, often accompanied by gamma rays, from the nuclei of an unstable isotope. As a result of this emission the radioactive isotope is converted (or decays) into the isotope of a different daughter element, which may or may not also be radioactive. Ultimately, as a result of one or more stages of radioactive decay, a stable, nonradioactive end product is formed.

<u>rainout</u> - the process of removal of particles of fallout from the air either by the formation of water droplets around the particles which then fall as rain, or by rain falling into the fallout cloud and "washing" the particles down to earth. Rainout does not affect fallout particles that are higher than about 10 km (33,000 ft).

<u>rem</u> - a unit of dose equivalent, numerically equal to the dose in rads multiplied by factors such as the quality factor, which takes into account the higher risk of late biological effects by certain radiations such as heavy ionizing particles (alphas, neutrons, protons) along their paths through cells of the body.

<u>RM (radiological monitor)</u> - the person who uses radiological instruments to (1) measure nuclear radiation intensities, (2) estimate the radiation exposure of shelter occupants, (3) find the places with the lowest nuclear radiation levels in a shelter, (4) advise on the improvement of radiation protection in a shelter, (5) advise when (and for how long) someone can go outside the shelter on short emergency trips, and (6) advise when to leave for longer trips, and when to leave permanently.

<u>roentgen (R)</u> - A unit of radiation exposure determined by the amount of ionization produced in air. Specifically, it has been defined as the quantity of radiation that will ionize dry air at zero degrees centigrade and standard atmospheric pressure to produce one electrostatic unit of electric charge of each sign, both positive and negative, in one cubic centimeter.

<u>shielding</u> - any material or obstruction which absorbs or attenuates radiation and thus protects personnel or materials from the radiation effects of a nuclear explosion. A moderately thick layer of any opaque material will provide satisfactory shielding from thermal radiation, but a considerable thickness of material of high density may be needed to protect adequately from nuclear radiation.

<u>skyshine</u> - radiation, particularly gamma rays from a nuclear explosion or from fallout, reaching a target from many directions, mostly from above, as a result of scattering by air.

<u>surface burst</u> - same as ground burst.

<u>survey meter</u> - an instrument used to measure the exposure rate in roentgens per hour at the location being metered.

tenth-value thickness - the thickness of a given material which will decrease the intensity of gamma radiation to one-tenth of the amount incident upon it. Two tenth-value thicknesses will reduce the intensity received by a factor of 10 x 10, or 100, and so on. The tenth-value thickness of a given material depends on the gamma-ray energy, but for radiation of a particular energy it is roughly inversely proportional to the density of the material.

thermonuclear - an adjective referring to the process in which very high temperatures are used to bring about the fusion of light nuclei, such as those of the hydrogen isotopes deuterium and tritium, with the accompanying liberation of energy. A thermonuclear bomb is a weapon in which part of the explosion energy results from thermonuclear fusion reactions. The high temperatures required are obtained in this case by means of a fission explosion.

x ray - a photon of high energy, or a stream of such photons, resulting from processes other than nuclear transformations.

yield - the total effective energy released in a nuclear explosion. It is usually expressed in terms of the equivalent tonnage of TNT that would be required to produce the same energy release in an explosion.

Appendix B

CONVERSION OF STANDARD TIME DESIGNATION TO TWENTY-FOUR HOUR TIME

In twenty-four hour time, used by airlines and military services, the first two numbers indicate the hours past midnight, starting from zero at midnight and going to 24 throughout the 24 hours of the day. The last two numbers in twenty-four hour time indicate the numbers past the hour.

Standard Time	24-hr Time
12:00 AM (midnight)	0000 hr or 2400 hr
12:30 AM	0030 hr
1:00 AM	0100 hr
1:30 AM	0130 hr
2:00 AM	0200 hr
3:00 AM	0300 hr
4:00 AM	0400 hr
5:00 AM	0500 hr
6:00 AM	0600 hr
7:00 AM	0700 hr
8:00 AM	0800 hr
9:00 AM	0900 hr
10:00 AM	1000 hr
11:00 AM	1100 hr
11:59 AM	1159 hr
12:00 PM (noon)	1200 hr
12:30 PM	1230 hr
1:00 PM	1300 hr
2:00 PM	1400 hr
3:00 PM	1500 hr
4:00 PM	1600 hr
5:00 PM	1700 hr
6:00 PM	1800 hr
7:00 PM	1900 hr
8:00 PM	2000 hr
9:00 PM	2100 hr
10:00 PM	2200 hr
11:00 PM	2300 hr
11:59 PM	2359 hr
12:00 AM (midnight)	2400 hr or 0000 hr

Appendix C

BIBLIOGRAPHY

The Effects of Nuclear Weapons, compiled and edited by Samuel Glasstone and Philip J. Dolan, prepared and published by the United States Department of Defense and the United States Department of Energy, 1977.

The Effects on Populations of Exposure to Low Levels of Ionizing Radiation: 1980, prepared by the Committee on the Biological Effects of Ionizing Radiations, published by the National Academy Press, Washington, D.C., 1980.

Nuclear War Survival Skills, ORNL-5037, Cresson H. Kearny, Oak Ridge National Laboratory, available from National Technical Information Service, U.S. Department of Commerce, 5285 Port Royal Road, Springfield, Virginia 22161, 1983; also available from Caroline House Publishers, Inc., 920 West Industrial Drive, Aurora, Illinois 60506, 1983.

Protection of the Thyroid Gland in the Event of Releases of Radioiodine, NCRP Report No. 55, published by the National Council on Radiation Protection and Measurements, 7910 Woodmont Avenue, Washington, D.C. 20014, 1977.

Radiological Factors Affecting Decision-Making in a Nuclear Attack, NCRP Report No. 42, published by the National Council on Radiation Protection and Measurements, 7910 Woodmont Avenue, Washington, D.C. 20014, 1974.

Sources and Effects of Ionizing Radiation, United Nations Scientific Committee on the Effects of Atomic Radiation, United Nations, 1977.

Appendix D

ACKNOWLEDGEMENT

INCORPORATION OF COMMENTS FROM SCIENTIFIC COMMITTEE 63, NCRP

In May, 1982, a letter was sent by the Federal Emergency Management Agency (FEMA) to the National Council on Radiation Protection and Measurements (NCRP) requesting comments from the NCRP Scientific Committee 63 (SC-63) on certain technical questions concerning material in this handbook. A letter from SC-63 in October, 1982, provided general comments in response to the following questions.

1. "Under what circumstances should people in a given fallout shelter be moved from one location to another in an attempt to reduce radiation exposures? There are subordinate questions that involve such things as a time-averaging technique of dose rate readings, and radiological instrument ranges and accuracies."

2. "What are the best ways to use radiation detection instruments to find radiation "leaks" into shelters and to verify the effectiveness of improvised corrective measures? A subordinate question relates to the utility and design of a special instrument for this purpose."

5. "Under what circumstances should philosophies such as "equal sharing of dose," or "women and children first," or "let's find a hero to volunteer" and the like be recommended? Might it be better to avoid all such discussion in preattack preparedness literature, leaving the consideration of such philosophies up to the people who would actually be involved in the circumstances that actually would occur should there be a nuclear attack on this country?"

Questions 3 and 4 were deferred by SC-63 for further consideration. The substance of the comments made by SC-63 in response to the questions quoted above has been incorporated, where applicable, in this handbook.

INDEX

(References are made by paragraph number)

☆U.S. GOVERNMENT PRINTING OFFICE: 1990 772-440/20150

RADIATION EXPOSURE RECORD

Name ______________________________

Home Address ______________________________

Social Security No. ______________________________

Shelter Address ______________________________

____

Name of Shelter

Unit Leader ______________________________

Rad. Sensitivity

Category ______________________________

Hour and Date	Added Exposure (R)	Total Exposure To Date	Comments

RADIATION EXPOSURE RECORD

Name ______________________________

Home Address ______________________________

Social Security No. ______________________________

Shelter Address ______________________________

____

Name of Shelter

Unit Leader ______________________________

Rad. Sensitivity

Category ______________________________

Hour and Date	Added Exposure (R)	Total Exposure To Date	Comments

Hour and Date	Added Exposure (R)	Total Exposure To Date	Comments

Hour and Date	Added Exposure (R)	Total Exposure To Date	Comments

RADIATION EXPOSURE RECORD

Name ______________________________

Home Address ______________________________

Social Security No. ______________________________

Shelter Address ______________________________

____

Name of Shelter

Unit Leader ______________________________

Rad. Sensitivity

Category ______________________________

Hour and Date	Added Exposure (R)	Total Exposure To Date	Comments

RADIATION EXPOSURE RECORD

Name ______________________________

Home Address ______________________________

Social Security No. ______________________________

Shelter Address ______________________________

____

Name of Shelter

Unit Leader ______________________________

Rad. Sensitivity

Category ______________________________

Hour and Date	Added Exposure (R)	Total Exposure To Date	Comments

Hour and Date	Added Exposure (R)	Total Exposure To Date	Comments

Hour and Date	Added Exposure (R)	Total Exposure To Date	Comments

RADIATION EXPOSURE RECORD

Name ______________________________

Home Address ______________________________

Social Security No. ______________________________

Shelter Address ______________________________

____

Name of Shelter
Unit Leader ______________________________
Rad. Sensitivity
Category ______________________________

Hour and Date	Added Exposure (R)	Total Exposure To Date	Comments

RADIATION EXPOSURE RECORD

Name ______________________________

Home Address ______________________________

Social Security No. ______________________________

Shelter Address ______________________________

____

Name of Shelter
Unit Leader ______________________________
Rad. Sensitivity
Category ______________________________

Hour and Date	Added Exposure (R)	Total Exposure To Date	Comments

Hour and Date	Added Exposure (R)	Total Exposure To Date	Comments

Hour and Date	Added Exposure (R)	Total Exposure To Date	Comments

Made in the USA
Coppell, TX
10 November 2026

11835572R00072